Deeper

Simon & Schuster

My Two-Year Odyssey
in Cyberspace

John Seabrook

SIMON & SCHUSTER
Rockefeller Center
1230 Avenue of the Americas
New York, NY 10020

SIMON & SCHUSTER and colophon are registered trademarks
of Simon & Schuster Inc.

Manufactured in the United States of America

10 9 8 7 6 5 4 3 2 1

Library of Congress Cataloging-in-Publication Data
Seabrook, John.
Deeper: my two-year odyssey in cyberspace/John Seabrook.
p. cm.
1. Internet (computer network)—Social aspects. I. Title.
ZA4201.S43 1997
302.23—dc20 96-36054 CIP
ISBN 0-684-80175-2

The author is grateful for permission to reprint the
following:

Fort Clark on the Missouri (February 1834),
engraving with aquatint, hand-colored, by Karl Bodmer.
(Courtesy Joslyn Art Museum, Omaha, Nebraska.
Gift of Enron Art Foundation.)

Engraving from a John Warner Barber drawing of the
southern view of Deerfield, Massachusetts, 1844.
(Massachusetts Historical Society © 1996.)

Lyrics from "Serial Killa" by Snoop Doggy Dogg,
from the album *Doggystyle* (1993).

Acknowledgments ➧ ➧

Seeds of this book were planted, more or less by accident, in some work I did for the *New Yorker*. Thanks to the people at the magazine who tended to the idea and helped it grow, especially to Deborah Garrison, for her brilliant editorial advice, and to Tina Brown, for encouraging me in the personal approach and for sending early pieces of the work out into the world with bells on them. Also thanks to the fact checkers, copy editors, and OK'ers who worked with me in the difficult business of writing accurately and plainly about computer technology, and for help in rendering Internet argot in English. Whatever errors remain are my own fault.

I am grateful to my agent, Joy Harris, the first to suggest that there was a book in this material, and to Alice Mayhew, for confirming Joy's suspicions, and for making me better than I thought I could be.

I am also grateful to friends who read the whole manuscript at various stages and offered valuable advice—Eric Schlosser, Dan Levy, Bev Talbot, Gerard Van Der Leun, Mark Boyer, John Homans, and especially my good friend Rob Buchanan, whose "Ugh!"'s and "Give Me a Break!"'s, written into the margins, saved me from at least some embarrassment.

And deepest gratitude to my wife, Lisa Reed, for reading everything, for listening patiently to my many accounts of amazing (to me) on-line dramas, and for always reminding me of who I am.

Finally, I am indebted to the many people of the Net for kind permissions to reprint e-mail and postings that served as signposts for me in my travels. A star-spangled banner reading IMHO (In My Humble Opinion) floats over these pages.

To my parents

Contents ◆ ◆

PART TWO ◆ ◆ East

I danced like a puppet, like a minion of Baal.

Michael Neubarth

Preface ◆ ◆

The Autobiography of Joe Homepage

This is the story of my life on-line. Although I did get out of the house once in a while to interview people for this book, my main strategy was to remain alone in my room, with my PowerBook on my desk, or sometimes on my lap, reading, lurking, e-mailing, chatting, posting, pointing and clicking, and observing the effects that all this time spent on-line had on my head. Attempting to explore a strange new land without leaving my room seemed like a good way of testing its most desirable feature: its promise to make the whole world only a mouse click away. Besides, in spite of its theoretically infinite space, cyberspace was actually pretty small in these early days, and most of the people I wanted to talk to showed up on my screen sooner or later anyway. Certain prominent sites on the Net, like the "com-priv" mailing list, or the Media Conference on the WELL, were like the "holes" that the mountain men used as refuges in the great American wilderness of the 1830s (Jackson's Hole, for example), where, if you stayed put long enough, you could see the whole West pass through.

A lot happened during my two years in my room; there was a lot to see on my screen. When I began, in September 1993, the Net was like a field, and in September 1995 the Web was like a state fair that had been set up in the middle of that field. When I logged on, the

Net appeared to be an extension of the PC revolution. Now the Web is looking more like the end of the PC revolution, and the start of the next phase. I started out as a total novice, a "newbie"—a word that in its suggestion of a first-year student in a military academy catches the situation about right. Now I'm Mr. E-mail, at least to my print friends. In the beginning I felt that special lightness of hope and possibility that new communications technologies seem to be uniquely capable of inspiring, a kind of spiritual feeling, which surprised me, as I don't like going to church. By the end I no longer felt that way about the technology, and I wondered whether the feeling had been an illusion, and whether I and countless others had in fact been duped by capitalists into requiring ever better, ever more expensive technology to maintain our "religion."

At first, I explored the Internet by day, and when it started to get dark, or when the weather turned ugly, I put up my high-tech geodesic dome for the night. In the morning I stuffed my home into its nylon sack and moved on, a technomad. There was no road or trail to follow, there was no landscape, no right or left, topography, solid terrain, gravity, or pole star. I navigated by metaphor. But any one metaphor for understanding the experience could only take you so far—the key was knowing at what point to dump it from your pack. I received generous assistance from other hikers I encountered on the trail. It's too bad Daniel Boone wasn't Al Gore's father, because the "Information Trail" would have been a better metaphor than the "Information Highway" for describing what this early part of my on-line experience was like. There was the spirit of the woods or the trail between us, or whatever that thing is that makes you feel friendlier toward your fellow man. It was nothing like that on interstate highways. When I stopped on a highway it was only for gas or to eat in a McDonald's. I couldn't remember the last time I talked about more than my meal with a stranger in any one of these places.

Much of my second year on-line was consumed with the search

for a place to call "home." It took a year, not only because there were so many places to look, but also because in looking I was trying to figure out what "home" meant, in an on-line context, and why I wanted one. In the real world, I knew I was approaching home when my brain subliminally recognized the olfactory pattern of the old spice warehouses in our part of Manhattan, from olive oil to coffee to nutmeg, walking east to west. I knew I was at home when I heard the sound of the front door closing behind me. My home was a way of keeping the world out, among other things. If, like me, you bought space in what used to be a warehouse, gutted it, and hired someone to turn it into a home, you invited the world into your life for a while, but as a result of that invasion you had walls and a threshold. Of course you never entirely shut the world out, just as we never seemed to be able to get rid of all the little packets of sugar that the Irish guys who built our loft brought along on the job with their coffee, but at least you had your privacy. It was like being inoculated with a little bit of the world, to make you better able to survive the whole world.

An on-line home, on the other hand, seemed more like a little hole you drilled into a wall of your real home to let the world in. E-mail, chat, conferencing, posting in newsgroups, and other forms of computer-mediated communication that I engaged in while sitting in front of my screen were a little like the coded tappings on the prison walls made by the prisoners in Arthur Koestler's book *Darkness at Noon.* An on-line home built for solitude didn't quite make sense, maybe because people tended to be alone when they sat in front of their screens. In going on-line, you made some of your personal space available to other people. That seemed to be partly the point of the exercise.

It was one of my purposes in my travels to give in to my feelings about this technology uncynically, and to report candidly how I felt at the time, even if subsequent experience taught me that my first impressions were misleading. When you start out on-line, it seems as though politics, ethics, and metaphysics—all the great disciplines of mankind—are reduced to their original elements, and are yours to remake again. You may find yourself pondering these fundamental questions for the first time, as you were too young or impressionable or unsure of yourself to think much about them the first time around. Why should individuals obey other individuals? What are the benefits of individual liberty, and what harm does that liberty do society as a whole? Why is honesty necessary? What is a neighborhood? What is a friend? Who am I?

Part One **West**

At length we gained the summit, and the long-expected valley of the Platte lay before us. We all drew rein, and, gathering in a knot on the crest of the hill, sat joyfully looking
down upon the prospect. It was right welcome; strange too,
and striking to the imagination, and yet it had not one pictur-
esque or beautiful feature; nor had it any of the features of
grandeur, other than its vast extent, its solitude and its wild-
ness. And yet stern and wild associations gave a singular
interest to the view; for here each man lives by the strength
of his arm and the valor of his heart. Here society is reduced
to its original elements, the whole fabric of art and conven-
tionality is struck rudely to pieces, and men find themselves
suddenly brought back to the wants and resources of their
original natures.

Francis Parkman, Jr., *The Oregon Trail,* 1849

Chapter One The Nerd Within

I. **"Okay, Go Ahead and Detach from the Mother Ship"** ◊ ◊ ◊

I saw a computer for the first time in September 1972, during my first month at boarding school. It sat in a windowless glassed-in corner in the basement of the main building, near the mailboxes and the school store. At the mid-morning break I would delay the possible disappointment of finding an empty mailbox by standing in front of the glass and looking in at the machine. In its size and industrial appearance it more resembled the nearby furnace than the sleek and beautiful screen I am staring into now. There was no monitor at all, just a loud teletype machine that spat out foaming billows of green and white computer paper. The floor was often littered with stiff blond cards that had little holes punched in them. These little holes were, according to Frenchy—short for Frenchy Le Pew, as the other new boys called him on account of his body odor—the language with which people talked to the machine.

Frenchy and I were two of thirty-five new boys who lived to-

gether in a large open room: a jungle of boys. Each boy had a
six-foot-high alcove, with a curtain to pull in front of it for privacy,
but there was open space all above, and other boys did not regard
those cloth-and-plywood squares as personal space. They felt free
to move in and out of your alcove at will, to bug you and horse
around with your belongings. Athletes were the rulers of our primitive society. Below the athletes in status were boys who were good
at making friends, or who were useful to the athletes in some way.
At the bottom were boys like Frenchy and me. Although I was full
of physical energy, it came flying out of my body in angry and
completely graceless ways. I was clever but did not seem to be good
at anything that counted for status. I don't think I had ever heard
the relatively new word "nerd" before arriving at school; I'm sure
I had not heard it used to describe me before. But within a few
weeks the evidence that I was a nerd was pretty conclusive.

Frenchy was one of three or four other new boys whose status
was even lower than mine. He was friendly toward me, for which I
was grateful, although I was careful to avoid letting the superior
boys see me hanging around with him if I could. Frenchy and I did
have a disturbing amount in common. For example, we both had
good aim. Aim is like a sport in which the will to succeed *can*
overcome lack of physical talent. We spent Sunday afternoons together at the riflery range, earning the small gold bars that denoted
rank on the National Rifle Association's marksmanship ladder. We
also loved board games. We had both been passionate jigsaw puzzle
builders earlier in life, and now we had graduated to war games
made by the Avalon Hill Company. These were very sophisticated
games, into which weather, terrain, troop strength, and various historical realities were all factored. They required some research,
much fanatical interpretation of rules, and days to complete. On
Saturdays we hid in an empty classroom on the top floor of the gym
and fought the battles of Jutland, Gettysburg, the Marne, and the

Bulge, with the windows open to let out the clouds of Frenchy's B.O., which let in the sounds of other boys cheering the athletes on various fields of glory.

I don't think that Frenchy liked being a nerd any more than I did, but he had known he was one for longer than I had, and he seemed resigned to his fate. I was never resigned to it. For me, everything depended on somehow acquiring the authority of the athletes. Studying how to accomplish that was to become the main focus of my energies during my four years at that school.

One Saturday after dinner we were sitting in Frenchy's alcove, talking in whispers, hoping to avoid persecution, although in that big open room it was impossible to hide, when Frenchy said, "Let's go play Lunar Landing." Lunar Landing was a computer game Frenchy had been trying to get me to play all fall. As he described it, the computer created a moonscape of mountains and craters and other hazards, as well as a landing module with thrusters and limited fuel. You typed instructions into the computer and it typed back your progress.

Because it was a Saturday night, the store would be closed, and nobody would be checking for mail, so I figured that I could go down to the computer room with Frenchy without much danger of being observed. We went downstairs. Frenchy got the punched cards out of his drawer, stretching his thumb and third finger as wide apart as possible, and picked up bunches at a time. In his wired, obsessive-sounding way—a characteristic I shared with him, but took pains to conceal—Frenchy explained that the computer could only speak computer language, which was a kind of symbolic language, sort of like the Boolean logic we were then studying in geometry class. The punched holes turned human concepts into the symbols that the computer could understand. You had to have the cards in exactly the right order—the procedure was crucial.

I could feel the formal, game-loving qualities of my intelligence flowing toward this new way of organizing the world. Perfection through skill and practice; control that could be achieved in silence, and alone. Authority, this was what I desired. But at the same time I felt a deep, irrational fear at the sight of the cards. The patterns the holes made in the cards reminded me of a nightmare I used to have when I was younger, especially if I went to sleep with a fever. In my dream I was floating above a field of white dots, and it was absolutely essential that all the dots remain white, for reasons that were never explained in the dream. When the first dot turned black I thought I must have not seen it right, because of course all the dots *had* to be white. Then I saw another dot turn black. Then a cancer of black dots was spreading from out of this contaminated portion, across the field of white dots, and there was nothing I could do to stop it, I could only watch. For some reason, this dream was terrifying.

Frenchy started feeding the cards into the computer. The machine sprang to life and the typewriter stuttered out <Hello>. We sat down at the keyboard and Frenchy said, "Okay, go ahead and detach from the mother ship."

"Like this?"

The teletype spat, <You are tilting to the left.>

"Fire thrusters," Frenchy said.

"Firing thrusters . . ."

<You have used six gallons of fuel.>

A loud rapping sound on the glass brought me back to earth abruptly. Two of the athletes were standing on the other side of the glass, laughing at us. One of them pressed his face up against the glass and said, "Neeeeeerrrrrddddds." The word rang in my ears like the time's-up buzzer in a TV game show. *Neeeeeerrrrrddddds.* Frenchy didn't seem to mind, or even to notice. He just went right ahead typing commands into the computer. "Stop typing!" I shouted. The boys remained outside the window, laughing. "This game is stupid!" I yelled, and walked out of the computer room and down the hall, past the mailboxes and the store, laughter curling around my ears.

I never went back inside that room. Fifteen years passed before I touched a computer again.

II. **Growth Units** ◊

Well, it wasn't only the nerd factor. I belong to the last generation of well-educated Americans for whom computer illiteracy was normal, even admirable. It was natural for a mind like mine to evolve independently of computers, in the great Loch Ness of the English major, so that by the time personal computers became widely available, in the mid-1980s, I had already conceived of the "computer world" as being completely separate from my world. While I was at Princeton University—where in 1995 almost all students had personal computers in their rooms, and where half the time in a two-hour seminar was routinely given over to discussing the e-mail that the students had written about the assigned reading—I saw no computers in anyone's room up through the year I graduated, 1981. At Oxford University, where I spent the next two years, computers were even scarcer. It wasn't simply that I lacked experience with computers— I'm saying that at some point, without my actually being aware that it was happening, I had developed a bias against computers. The computer was a symbol of a mechanized, passionless approach to life that it was a literary person's duty to resist.

Where did this attitude come from? Certainly not from my father or from his father, both of whom were engineers and technologists, the sort of pragmatic, problem-solving Americans whom Alexis de

Tocqueville met on his tour of the United States in the 1830s, for whom "every new way of getting wealth more quickly, every machine which lessens work, every means of diminishing the costs of production, every invention which makes pleasures easier or greater, seems the most magnificent accomplishment of the human mind." My grandfather was a South Jersey truck farmer who built his fortune on frozen vegetables, getting in early on that particular technology's revolutionary effect on the fresh-vegetable business. My father, on taking over the business in the 1950s, expanded into such modern miracles of domestic science as frozen concentrated orange juice and TV dinners, and was responsible, he claimed, for the boil-in-bag technique of cooking frozen creamed spinach. At Seabrook Farms, technology was relentlessly applied to every aspect of agriculture. Temperature, sunlight, water, and seed were all taken into account and factored together as "growth units," which supposedly allowed farmer-technologists to calculate optimum ripeness to the hour. Harvest was a highly organized military assault on vegetables—the agricultural equivalent of D-Day. Although my father and his father were proud of their attachment to the earth, I think they would have been almost as happy disposing of industrial waste, say, as they were raising vegetables. The goal was not to achieve harmony with nature but victory over nature. Anything that could not be factored into "growth units" (eerily reminiscent of the "sunshine units" promoted by the Atomic Energy Commission), such as poetry, religion, or emotion, did not interest my paternal ancestors very much.

I sometimes argued with my father about nuclear energy. As in all our arguments, I was the "liberal," an idealist whose hypocrisies and received ideas my father was delighted to expose. He claimed to have once seen a car with two bumperstickers on it, one saying "Split Logs, Not Atoms," and the other saying "Stop Acid Rain," and he liked to bring this up in the midst of these discussions. Couldn't the poor granola-crunching, tree-hugging liberals understand that burning fossil fuels *causes* acid rain? I never beat him at these arguments, but I almost always retired from the field smug in the knowledge that I was right. Later, alone in my room, I would triumph over him in my head, putting his arguments to rout.

Nevertheless, part of me had unconsciously imbibed my father's legacy: I believed in progress. Quaint nineteenth-century notions of social progress through mechanical innovation inhabited my soul in ways I was hardly aware of. The belief expressed itself in my love of gadgetry, of gear, of science—but although the impulse found expression in the material world, it was actually sort of like my religion. Faith in progress was my way of believing that there was meaning and order in the universe. As I got older my faith in the gradual improvement of all things expanded and deepened into a love of the idea of "universal history," beautifully expressed by Hegel, and then by Marx—the idea that human history was the story of progress in the condition of mankind, tending toward a state of perfection. I have never believed in any other philosophy so devoutly as I believed in Hegel's *Philosophy of History,* a book I read at age twenty-two in the Radcliffe Camera, lower level, a paradise of books where I spent a large part of my two years at Oxford. Hegel and Marx lead me to Concordet and Comte and Herbert Spencer. These were writers who had actually discovered the scientific *laws* of progress. I particularly liked Spencer's idea that evil is not inherent in mankind, but instead is a condition caused by man's imperfect relationship to the world—a condition that technological innovation will gradually ameliorate.

The existence of progress meant it was *logical* to believe that everything happens for the best, and that was as close to an act of faith as my skeptical Yankee mind would tolerate. So when I came across the eighteenth-century Puritan divine Jonathan Edwards asking in one of his sermons, "Does God make the world restless, to move and revolve in all its parts, to make no progress, to labor with motions so mighty and vast, only to come to the same place again, to be just where it was before?" I naturally thought the question was rhetorical, and that the obvious answer was no, God does not —God makes the world restless for a reason! Things *had* to be getting better, and I was willing to do my part to make sure that they did, because if they weren't getting better—well, that was the thing that could not be. Behind that idea lay the fear, the darkness, the black dots, the worm.

The idea of progress was transmitted to me by my family—for without technological innovation, would I not still be down on my knees in the dirt, as my grandfather had been, weeding onions in the heat and mosquitoes?—and by the two-bean blend of Calvinism and Puritanism that they served in Presbyterian Sunday school, and by various gym teachers ("No pain, no gain!") and coaches (the sport of rowing, which I discovered in my second year of boarding school, and which would prove to be my ticket out of nerd-dom and into the sunshine world of the athlete-gods, was the Calvinist sport par excellence: if you worked the hardest, you would be the best). Although I never attended a single religous service at Princeton—a hotbed of "awakened" Puritanism, when the university was founded, in 1746—the secular version of the religion was strong in the English Department, where it was called "tradition" and "influence," and above all else "hidden meaning," and passed through F. R. Leavis and John Crowe Ransom and Allen Tate, not through Increase Mather and Jonathan Edwards, as it had been in the old days. Had I studied more religion in college I might have been more aware of the peculiar religiosity of literature studies than I was.

I almost never talked about progress with other believers, although I noticed that there were a lot of us around, and that we tended to occupy positions of authority—worldly success being a possible (although by no means definite!) sign of our personal salvation. I kept my light under a bushel, for in the 1970s it seemed sort of stupid to believe in progress—especially in social progress through mechanical invention. The prospect of world annihilation by means of the marvelous progressive technology of atom-splitting made it very clear that a technology invented by good people for good reasons can also be used by evil people for evil reasons, and that unless mankind makes corresponding moral progress, social progress through technological innovation is impossible. As Lewis Mumford put it in 1970:

> By turns the steamboat, the railroad, the postal system, the electric telegraph, the airplane, have been described as instruments that would transcend local weaknesses, redress inequalities of the natural and cultural resources, and lead to a worldwide political unity—"the parliament of man, the federation of the world." Once technical unification was established, human solidarity, "progressive" minds believed, would follow. In the course of two centuries, these hopes have been discredited. As the technical gains have been consolidated, moral disruptions, antagonisms, and collective massacres have become more flagrant, not in local conflicts alone but on a global scale.

Landscape blackened by pollution, water poisoned by petrochemicals, lungs destroyed by asbestos, babies deformed by thalidomide, and above all else, a world in the shadow of nuclear Armageddon —this was the legacy of technological progress.

But to those of us who derived our ideas from the New England wing of the liberal mind, where the practical genius of the early Yankee machinists was steeped in the philosophy of progress, this long winter of forced hostility to the progressive promises of technology was hard to endure. I accepted it, but it didn't make sense to me. Why shouldn't technology make things better? Wasn't humanity better off for the pencil? For the polio vaccine? For books? For electric light? How could you not believe in progress through technology? You could see it.

III. **The Flying Squirrel Effect in Reverse** ◊ ◊ ◊ ◊ ◊ ◊ ◊ ◊ ◊ ◊

In 1987 I moved into a railroad apartment in a small federal building in Greenwich Village, a dump with potential that was not realized during my time in it. The only other apartment in the building was occupied by a young man named Mark Boyer. On my second day in my new place he and his girlfriend, Lindsay, invited me down for a drink. They were collectors and hobbyists, and their place was packed with evidence of their enthusiasms: antique toasters, tropical fish, forms for dress- and hat-making, Indian cookbooks, pet budgies. Gesturing toward a pile of recording equipment, Mark explained he was in the process of setting up a business called "Record-a-Pet International"—for $19.95 he would digitize a recording of your pet's voice and remix it to the tune of "Rockin' Robin," "The Blue Danube," or "Love Me Do."

"What's that?" I asked, pointing to a putty-colored box that sat among the jumble of wires and reels.

"That's a Mac," Mark said. With a cigarette in the side of his mouth, he arched his eyebrows and asked, "Wanna see?"

A lot had changed in fifteen years. The punched cards I remembered from Frenchy's drawer had been replaced by floppy disks, and these had the instructions for the machine—which Mark called "software"—encoded inside of them. Also, this computer had the capacity to store the basic operating instructions, so you didn't have to program the machine every time you used it. I had seen word-processing machines, which made awful-looking green or orange glowing words on a screen, but Mark's monitor was different. It could make pictures—files, pages, rulers, and a trash can, and other graphical doohickeys representing programs, that Mark refered to as "icons." Sitting down at the keyboard, Mark explained that in the Macintosh system, which used a "graphical user interface"

(GUI), you operated the computer the same way you "operated" your desktop: by nesting documents inside folders and arranging them on your "desk"—the two-dimensional plane of the monitor itself. When you wanted to open a document you pointed a cursor at it by moving a plastic thing called a mouse, and then "double-clicked." (My God, this seems like ancient history now.) There were no numbers, no mathematical symbols to input into the machine; the Macintosh software hid all the symbolic language beneath a layer of illusion.

Mark was extremely enthusiastic about his Mac. He took me for a spin through some of its wonders. He extolled the Mac's virtues over those of the IBM PC, which used an inferior operating system known as DOS, a command-line interface that was the leading product of someone named Bill Gates, whose company, Microsoft, produced it.

"Well," I said, getting up from the computer, "maybe it's time for me to get a Mac myself."

"Definitely it's time," Mark said. "It's long past time. Don't put it off another day." Then he lit another cigarette for himself, detaching a single match from the clump with one hand, twisting it around the pack and striking it with his thumb—all in one smooth hepcat gesture. I didn't know it then, but Mark was the very first "digital guy" I had ever met.

I managed to put it off for another year, but in 1988 I bought a Mac of my own and started using it to write magazine articles, which was my profession. I learned to "Cut" and to "Paste." I liked using the mouse. It was like an electronic trowel with which you planted and weeded words. I learned that the brain of the computer was the microprocessor, or the "chip"—the "fingernail-sized" sliver of silicon on which millions of tiny transistors were crammed (the fingernail seemed to be a standard unit of measurement in electronics). I learned that the computer also had a hard drive and a memory. Memory, or RAM, I could understand because it sounded like human memory, even though it turns out that computer memory is more like "attention" than memory, and the "hard drive" was like an expression you might see in a hotrod magazine: the engine or something. The desktop metaphor—my first metaphor!—helped me understand how the software was organized and where everything was. It was an early, two-dimensional version of cyberspace. The desktop metaphor did not, in fact, always correspond to the way I used my real desktop (a jumble of books and overlapping mounds of paper, with scissors, pencils, tape, and erasers scattered around the edges of the pile), and in those moments I could feel the intuitive element in my thought process confronting the Boolean logic of computers, and squirming like a slug under a pinch of salt. But I found that after I had been using my Mac for a while, my real desktop got neater and more logically organized—more like the metaphorical desktop on my computer screen. I actually bought a filing cabinet and began trying to organize my papers into folders, which I had never done before. Having begun by using my brain as a metaphor to understand my computer, I was now using the computer as a metaphor to understand my brain. After an efficient morning's work, I would think with satisfaction of my mental processor whirring, my cerebral software distributing information throughout my cortex. Had I been going around the streets thinking of my brain as a telephone, a radio, a television, or any other household appliance I can think of, I would probably have been picked up and taken to the hospital. But a computer seemed to be an acceptable appliance to have implanted on one's shoulders. It was a "thinking machine."

Writing on a computer made the arrangement of the parts of an article much easier, which made writing more of a formal exercise for me, a game. The ease of typing on the electronic keyboard made it possible to write almost at the speed of thought, in riffs, and the

"Cut" and "Paste" tools made composition above all else a matter of arranging all the riffs so that they fit, each new block of words causing displacement in the blocks alongside it, which I would then smooth over with my electronic trowel. Sometimes, when I was in the middle of writing a piece, my editor would call and say, "Have you printed it out and looked at it yet? You haven't printed it out yet? How do you know how good it is until you print it out?" And although this didn't seem logical to me—the words themselves were the same, after all, whether on paper or on the screen—it did seem to be true. Sentences that had seemed to flow smoothly and logically when read in glowing pixels appeared weak and disjointed in cold type. I wondered if there were something about the light coming from the screen that dazzled my critical judgment and lit up some atavistic primate center of my brain, the same place that had first been dazzled by fire.

The computer had made my work easier, even better, I believed, and I loved it for that, but none of the routine disasters of daily life, not even the chance of the cat falling off the roof a second time— my cat had been miraculously saved from her first "feline high-rise" by what the vet called "the flying squirrel effect" (the loose pouches of skin under the cat's legs caught the air and caused her to glide rather than plummet)—worried me as much as the prospect of losing my work because of my computer "crashing." Fire, flood, or theft could destroy my work, but there was no chance that my typewritten pages of writing were going to flash me a message saying "Serious Disk Error," and then disappear because of some inexplicable glitch in the paper or ink. With my computer, there was always the chance of the inexplicable calamity happening, like the flying squirrel effect in reverse. Once in a while something would go wrong in the communication between the computer and the printer, and the latter would spit out code instead of words, like some kind of enraged machine rant—

426161c2069732067726561742c204261616c20697320676f6f
6420616e64207765207468616e6b2068696d20666f72206f75
7220666f6f642e204261616c2069732067726561742c204261
616c20697320676f6f6f6420616e64207765207468616e6b2068
696d20666f72206f757220666f6f642e204261616c2069732067
726561742c204261616c20697320676f6f6f6420616e64207765
207468616e6b2068696d20666f72206f757220666f6f642e204
261616c2069732067726561742c204261616c20697320676f6f
6f6420616e64207765207468616e6b2068696d20666f72206f7
57220666f6f642e204261616c2069732067726561742c20426
1616c20697320676f6f6f6420616e64207765207468616e6b2068
696d20666f7226f757220666f6f642e204261616c2069732067
726561742c204261616c20697320676f6f6f6420616e6420776520

When that happened it was like the part in the science fiction movie when the android, who you had thought was human, and whom you had fallen in love with, cuts herself, and you see the circuitry underneath the skin. Words have warmth and old associations and are pleasantly worn and sculpted by many centuries of use. Code by comparison seemed cold, dead, and cheerless.

IV. **Prophets of More** ◊ ◊ ◊ ◊ ◊ ◊ ◊ ◊ ◊ ◊ ◊ ◊ ◊ ◊ ◊ ◊ ◊ ◊ ◊

But just as I became used to the idea that my computer was a typewriter, it changed into a telephone. At first I wasn't happy. I wanted my computer to stay the way it was. This is the problem with progress—once you commit yourself to believing in it, you have to keep upgrading your stuff. I spent the first several years of the 1990s trying to ignore this new enthusiasm for computers as communications machines, hoping that it would go away, and that the time I did not invest in learning more about computers would thus be time wisely saved. My only concession to technological progress was to replace my old Mac SE with a black-and-white PowerBook, a laptop that was, thanks to Moore's Law, now a more powerful machine than my old desktop model.

Moore's Law was named after Gordon Moore, one of the founders of the Intel Corporation, the leading manufacturer of microprocessors. The law states simply that the growth in the power of microprocessors—that is, the number of transistors that will fit on a silicon chip—will double every eighteen months, or that the price will fall by half. That meant that in twenty years what now took a year of computing would take fifteen minutes. Although no one knew for sure what society would do with this power, nor even whether or not it was desirable, it would exist whether we wanted it or not. No natural calamity or political upheaval short of worldwide anarchy was powerful enough to stop it.

To a believer in progress, Moore's Law was a kind of revelation. Every tool is to some extent an artifact of progress. Thousands of years of human consciousness of the problems that a hammer solves have flowed around its peen, tongs, and shaft, with succeeding minds adding improvements to the previous inventors, smoothing and shaping all the parts of the tool into the hammer I hold in my hand today. But in the case of microprocessors, the progress was much more dramatic. Twice as fast every eighteen months! Growing exponentially! Bringing computing power that was previously only available to rich corporations to the desks and laps of ordinary citizens! It was wonderful!

George Gilder was one of the leading advocates of the utopian possibilities of Moore's Law:

> The new age of intelligent machines will . . . relieve man of much of his most onerous and unsatisfying work. It will extend his lifespan

and enrich his perceptual reach. It will enlarge his freedom and his global command. It will diminish despots and exploiters. It may even improve music and philosophy. Overthrowing matter, humanity also escapes from the traps and compulsions of pleasure into a higher morality of spirit.

In Gilder I recognized one of my coreligionists. His work had it all: the (apparently) logical proof of progress through technology, as well as the dread of the refining fire that everyone would have to go through in order to enjoy that progress—revenge of the nerds meets Jesus' parable of the workers in the vineyard. He had the impulse toward mysticism but distaste for mystical-sounding language, preferring instead words like mips, CPU's, rasters, and ATM switches. The epigraph of his book *Microcosm*—"Listen to the technology and find out what it is telling you," an injunction Gilder had picked up from the physicist Carver Mead and turned into a principle in his system of technological determinism—was a classic modern restatement of the old Calvinist idea of the "hidden God," the God who only reveals himself to the faithful through "illuminations." Only the hardworking, patient, and keen—the elect—would be able to hear what the technology was saying. Then, because the technology had not developed the technology of speech yet, those fortunate souls would be entitled to speak for it.

Of all these souls, Bill Gates was the most fortunate. In him, the technology spoke the loudest. His power in the software industry was metaphorically comparable to John D. Rockefeller's control over oil at the end of the last century, but not literally comparable, because oil was a finite resource, whereas software was infinite, or at least as infinite as human intelligence. To what extent Gates was a kind of historical accident, and to what extent he was the first person to imagine software as a shrink-wrapped commodity, and was therefore a visionary, was a good question to ask if you found yourself seated next to a computer-industry executive at a dinner party. It was often said by Gates's detractors that he had never invented anything, and that was true in a sense. Windows, which slowly replaced DOS as the standard operating system in IBM and compatible computers, was a knockoff of the GUI developed at Apple (Apple, in turn, had taken the idea of the desktop metaphor from Xerox, where it was pioneered). But you could say that Henry Ford never invented anything either. Like Ford, Gates was "technical"—he knew enough about the technology he was selling to understand how it worked—but his real genius was as a businessman: he had brilliantly and ruthlessly exploited his early position in the software business to achieve world dominance in the industry. When the Model T appeared, in 1908, it was by no means the best car on the road, but it worked well enough, and it was affordable and easy to produce, and Ford stayed with it. Likewise, Windows was by no means the best operating system available when it appeared—it was a "kludge" (an inelegant patch on top of DOS); it was "buggy" (it had glitches); and it was a "memory hog" (it used up lots of RAM). But it worked well enough, and Gates had stayed with it. Most users found Apple's GUI much easier to use than Microsoft's, but the number of computers running DOS and Windows was so much larger than the number running Apple's operating system (about nine times larger) that other software developers were much more inclined to make applications for Microsoft.

As a writer, I was slightly afraid of Gates and everything he represented. But as a believer in progress, he appealed to me. Moreover, the nerd in me, so long concealed within a crawl space underneath my personality, was emboldened by Gates's success to poke out its own unattractive little snout. Who would have thought that the lowest creatures in my prep school cosmology, the nerds, would be in charge twenty years later, and that the establishment types—the people I had invested considerable energy in emulating—would now effectively be worshipping the nerds, and that, moreover, this state of affairs would by and large be celebrated, and not regarded with horror?

I saw a photograph of Gates and his boyhood friend Paul Allen, with whom he would later cofound Microsoft, taken at a computer terminal in Lakeside School, a prep school in Seattle, in 1968. The fifteen-year-old Allen was seated at the terminal, which was strikingly like the terminal I remembered from *my* prep school; the thirteen-year-old Gates was standing to one side and staring at the machine with a kind of feral alertness and deep fascination. That picture brought back the memory of Frenchy and me in the computer room so vividly that I could almost smell him. Alas, poor Frenchy. I had done him wrong. He had offered me friendship, and I had paid him back by washing my hands of him as soon as I managed to acquire less nerdy friends.

"I realized later part of the appeal was that here was an enormous, expensive, grown-up machine and we, the kids, could control it," Gates later recalled in his book *The Road Ahead,* describing his early encounters with a computer. "We were too young to drive or to do any of the other fun-seeming adult activities, but we could give this big machine orders and it would always obey." Yes, that was it. It was that desire for control, that same obsessive quality that I remembered feeling in the computer room with Frenchy. As a boy Gates had been obsessive about improving aspects of himself he didn't like. "He was always upset about his little toe curling in, so he'd work on it. He'd spend time holding it out so he'd have a straight toe," his sister Kristi said in *Gates,* a 1993 biography. Gates used to try to impress his sisters by jumping out of a trash can, and he still occasionally jumped over his office chair from a standstill. Sometimes, on his way to a business meeting, he would suddenly jump up and try to touch as high as he could on a wall, or to touch higher than the spot he touched last time, but he said in *Gates,* "I don't jump spontaneously the way I used to, in the early years of the company . . . or even in a meeting. . . . Now the jumping is not that common." However, he was planning a trampoline room in the famous house he was building on the shores of Lake Washington, outside Seattle.

I wrote Gates a letter in Microsoft Word (would he be able to tell?):

29 July, 1993

Mr. Bill Gates
Chairman, The Microsoft Corporation
1 Microsoft Way
Redmond, Washington 98052

Dear Mr. Gates:

The *New Yorker* magazine would like to do a 15,000-word profile of you and your company, Microsoft. Tina Brown, our editor, has asked me, a staff writer who covers technology in society, to write the story. We'd like to focus on the emerging digital media industry and Microsoft's role in it. The angle would be something like this—the future of the computer is with Microsoft; let's go visit Bill Gates and try to figure out what the future will be.

Obviously it would be easier to write such a story if I could visit you and your company. Could I come out to Redmond and get a tour of Microsoft and talk to key people about Microsoft's plans in multimedia? [I wasn't sure what to call "the future," and I hadn't heard of the Internet yet, so I just called it "multimedia."] Could I sit down with you? Most of my questions for you are about the future —how will this technology change our lives? Also, could I be an observer on one occasion when you interact with other people—say, a business meeting?

Thank you and I look forward to hearing from you.

Yours truly,
John Seabrook

I sent my letter via Federal Express. After a short delay I was told that "Bill" had granted me an hour of his "face time" on a date two and a half months away. In the meantime I was invited to Redmond, where I could talk to other Microsoft executives, and have a look around the Microsoft campus.

The Microsoft campus resembled a college campus, including playing fields and employees in T-shirts and jeans who weren't much older than college students. Unlike many colleges, there was no attempt at Microsoft to inspire lofty thoughts with architecture: all the buildings were the same dull, two-story, eight-sided structures. (The architecture was so monotonous that the only way to find your way around campus was to follow the numbers of the buildings.) Nowhere on earth did more millionaires and billionaires go to work every day than here (about three thousand of the fifteen thousand employees were worth at least $1 million in Microsoft stock, and dozens were worth in excess of $100 million), but the campus was in no respect worldly. Workers spent much of their day staring into large computer monitors, occasionally exploding into a rapid fingering of keys. Diet soft-drink cans and cardboard latte cups collected on their desks. Designing software was a sort of intellectual handiwork. Operating systems, the most monumental of all software constructions, were like medieval cathedrals: thousands of laborers worked for years on small parts of them, crafting 1's and 0's into patterns that control switches inside microprocessors. The platonic nature of software—it was invisible, weightless, and odorless; it didn't exist in the physical world—seemed to determine much of the culture that surrounded it.

Although I had been anxious about interviewing people at Microsoft—I worried that it would be like trying to read the Microsoft Word manual, which always seemed to make two puncture wounds on my intellect from which my energy rapidly bled—in fact my trip to the campus turned out to be very useful. It seemed possible to communicate with the software developers on the level of metaphor, without ever descending to the level of code. Metaphors were the traveling salesmen that software makers sent out into the world to sell their line of merchandise. Metaphor had about the same significance in selling software that the jingle had in selling soft drinks. The trick was to come up with a metaphor that not only helped explain what the software did, but also made it sound desirable. For example, the word "smart" was a kind of metaphor widely employed at Microsoft. Smart meant something that used software, that had "intelligence" built into it, but of course it was also desirable to *be* "smart." The people I interviewed at Microsoft often referred to each other as "smart," or "supersmart," or "one of the smartest people you'll meet around here," and Gates was believed to be the smartest person of all. "Bill is just smarter than everyone else," Mike Maples, then an executive vice-president, told me. "There are probably more smart people per square foot right here than anywhere else in the world, but Bill is just smarter."

We discussed software programs called "intelligent agents," which were then beginning to be developed at Microsoft, but which would ultimately be your digital representative on the information highway, filtering the storm of information blowing through the wires, and grabbing the useful stuff. If, for example, you wanted to keep abreast of articles and discussions about global warming, you could program your agent to search for those keywords. Being "smart," these agents would learn about you as you used them. You might keep the same agent throughout your life, and perhaps will it to your favorite grandchild at the end of your life: your agent would be like the black box recovered from the wrecked cockpit that is your head. If you could develop an agent that was as smart as a Harvard graduate, people might be willing to pay as much for it as a Harvard education. In the near term, the plan was to develop agents that would do simple things like pop up in the corner of your television set and tell you that the president of Haiti was on Channel 4, because it knew you were interested in Haitian politics. As the agents became steadily more intelligent, they would begin to replace more of the functions of human intelligent agents—stockbrokers, postal workers, travel agents, librarians, and reporters like me.

At first I could not figure out what seemed unusual about the Microsoft offices: it was that the phones hardly ever rang. In the silence you could hear the chatter of fingers over computer keys and the little belches or squeaks that the computers sometimes emitted, indicating incoming electronic mail. Mike Maples told me that employees sent a total of 200 million e-mail messages a month, but when I got home and did some math (a rare activity for me)—that worked out to about six hundred messages a day, each—I figured that couldn't be right. Anyway, employees were heavy users of e-mail. At a lunchtime cookout in a courtyard, I heard several people start conversations by asking about "e-mail from Bill." "Did you get e-mail from Bill today?" I heard one young man ask. Another man said, "Did you see billg's mail?" Because most employees at Microsoft were much more likely to encounter their boss in digital form than in person, many had fallen into the habit of referring to him as <billg>, his e-mail handle, rather than Bill. (No one called him Mr. Gates.) Later that day, while I was talking with a programmer in his office, his computer made a little squeak, and it turned out to be e-mail from Bill. Someone else said, "Hey, that's a good idea, I'm going to e-mail Bill about that." Gates was not on campus at that time; in fact he was on the other side of the world, in Africa, touring the valley where the oldest human skeleton, Lucy, was discovered—but I had the sense that he was somehow present, distributed through the local area network, flying around the walls and popping into people's computers.

Shortly after returning from that trip, I was sitting in my study in front of my screen on an afternoon in late September (by now I was married to Lisa Reed and living farther downtown, in the loft we had fixed up) when it occurred to me that, in theory at least, I too could send e-mail to Bill. Since Gates's face time was so limited, maybe I could spend some time in his digital presence. What would that be like? I had no idea. Back then—so many screen years ago! —I did not have a clear idea of what e-mail was, other than it was something people did between computers. I did know that you needed a modem to do it. So, without allowing myself time to reconsider, I wheeled my bicycle into the elevator, went down to the street, and started riding down to J&R Computer World, a high-tech emporium in lower Manhattan, to buy a modem.

It was a fine Indian Summer day in New York City. As I rode through the streets I did this thing in my head that I had picked up from the *Terminator* movies. Approaching an intersection at speed with a red light against me, trying to anticipate an opening that might develop a second later in the stream of pedestrians and traffic, I imagined a little computer screen inside my brain, rapidly crunching velocities and vectors and the tiny nuances in people's body language, then flashing the results to my hands and feet, which then sent them on to the bike.

On that day in J&R Computer World there was a funny, chemical smell. Maybe it was the smell of nickel-cadmium batteries, or it may actually have been the smell of testosterone, but it seemed to be having an amphetamine-like effect on the customers: everyone was excited. Pin-striped circles of businessmen were gathered around the latest laptops, looking at them in the way that primates look when their curiosity is aroused—hushed, watchful, heads slightly on one side, wheels spinning behind the eyes. That awful-sounding series of squawks and buzzes that modems emit when they connect to each other seemed to be coming from everywhere in the store.

As I walked through the row of machines I started to feel as though I was carrying this black knapsack of anxiety. It was stuffed full of my ignorance about computers and software, and it was so heavy that I was having a hard time catching my breath. Here in this store was another in a series of recent signs that my education, and my profession—this making of small nuanced judgments about words—were useless in the modern world. I had blown it. I had not been "smart" enough. Yes, progress existed—here was a storeful of it—but the bad news was that it was leaving me behind. I was becoming an "information have-not."

I stopped one of the salesmen as he was hurrying across the floor and told him I wanted to buy a modem for a Macintosh.

"What baud rate?" he asked, speaking rapidly.

"What are my choices?"

"The fastest is fourteen four."

Pausing, head to one side, as though calculating whether this would suit my needs, I said, "Yeeeesss, that sounds about right."

The salesman led me to the modem, which cost $235, then to another part of the store, where he placed in my hands a box of software from an "on-line service provider" called CompuServe, which he said I would need in order to send e-mail. The booklike shape of the CompuServe box, as well as the illustrations on the outside—some battling chess pieces in a dreamlike, cobalt-colored, half-human and half–make-believe attitudes—reminded me of the Avalon Hill war games that Frenchy and I used to play so obsessively. This made the anxiety knapsack feel a bit lighter.

"Pay up front," the salesman said, and hurried away. Within twenty minutes, a shopping bag full of new information technology looped over his handlebars, the Terminator was on his way home.

V. E-mail from Bill ◊

Back at my desk I plugged in the modem and installed the software that came with it. This seemed to go smoothly. Then I popped in the CompuServe floppy disk, installed the application, and clicked the "Connect" button. My modem made the squawks and buzzes. Once connected, I followed the on-line sign-up instructions. Within a half hour, I seemed to be on-line. It wasn't too much harder than taking money out of a cash machine for the first time, except that in this case I was the cash machine and CompuServe was taking money out of me.

Now I needed Gates's e-mail address. In the course of my reporting I had met a man who was working on a project for Microsoft, and he had given me a business card that had *his* e-mail address on it. It was the only e-mail address I knew, besides my own. So I dug up the card and sent the man my virgin e-mail message, asking him if he happened to have Bill Gates's e-mail address. In a couple of hours he sent it back to me. I could have almost guessed it: billg@microsoft.com.

Then I wrote my second e-mail message:

From: <73124.1524@compuserve.com>

Dear Bill,

I am the guy who is writing the article about you for the *New Yorker*. It occurs to me that we ought to be able to do some of the work through e-mail. Which raises this fascinating question —What kind of understanding of another person can e-mail give you? . . .

You could begin by telling me what you think is unique about e-mail as a form of communication.

John

I hit the <return> key on my keyboard and the message disappeared from my screen. Did I send it, or did I just delete the message? I walked out to the kitchen to get a drink of water and played with the cat for a while, then came back and sat down in front of my computer again. Thinking to myself that I was probably wasting money, I nevertheless logged on again and saw a little "1" under the envelope icon, indicating I had a new message. I double-clicked on the icon and saw on the screen:

From: Bill Gates <billg@microsoft.com>

OK, let me know if you get this e-mail.

E-mail is a unique communication vehicle for a lot of reasons. However email is not a substitute for direct interaction.

There are people who I have corresponded with on e-mail for months before actually meeting them—people at work and otherwise. If someone isn't saying something of interest it's easier to not respond to their mail than it is not to answer the phone. In fact I give out my home phone number to almost no one but my e-mail address is known very broadly. I am the only person who reads my e-mail so no one has to worry about embarrassing themselves or going around people when they send a message. Our e-mail is completely secure.

E-mail helps out with other types of communication. It allows you to exchange a lot of information in advance of a meeting and make the meeting far far more valuable.

E-mail is not a good way to get mad at someone since you can't interact. You can send friendly messages very easily since those are harder to misinterpret.

My God! This thing actually works! This e-mail from Bill blew the fog away from my screen and made me see clearly that my previously unnetworked Mac was like a car that I had been sitting in for five years, fiddling with the radio and the windshield wipers, revving the engine, turning on the heat when it got cold, but never realizing that the point of the machine was to *drive* it.

Since this moment was the real beginning of my life on-line, of which these pages are a record, I should have said something memorable like, "We have struck a new power!" which was what one of the pioneers of the internal combustion engine supposedly said to another as they were turning the crank of a prototype of that machine, when suddenly the engine sprang into life and tore the crank from their hands. Or at least, "I got a response!" since in the end so much of what was attractive about the world beyond the screen would come down to this: recognition, a voice speaking from out of the howling void and saying, Yes, you are heard.

But instead I said only "I got e-mail from Bill!" and then walked down toward the kitchen saying, "I'm on-line," "I'm wired," "I got e-mail from Bill," in the jokey, ironic voice of that copyroom guy on "Saturday Night Live."

Chapter Two <billg>, BMOC

I. The Would Have ◊

So that was the first day. Did I draw back at this point and ask myself, Do I want to venture any deeper into the world beyond the screen? Is this a looking-glass world, in which my assumptions about the universe and my place in it are apt to come in for some pretty harsh peer review, perhaps leading to a kind of revenge of the nerds for my sins against the Frenchys of the world? No, I did not. Actually, like millions of other newcomers who suddenly found themselves chatting with a stranger in India, or downloading free software, or cutting and pasting swatches of Melville's prose into their term papers (easier than typing it in) I thought only, This is *great!* I needed an interview with Bill Gates, I have been told that I'll have to wait two and a half months to get one, and now, in less than twenty minutes on-line, I have found a solution to my problem. (So it often begins—you put off going on-line until the day comes when you need one specific thing, something you can't find in the library or on the telephone, and to your amazement find the thing on-line, and you're hooked.) I also felt a sense of relief. Maybe I was not going to be left behind by the information age after all. . . . In fact, I had to struggle to keep myself from firing back to <billg> the first response that came into my head. A pause seemed like dead air somehow. (I guess I was thinking of a telephone conversation.)

I mention all of this only as a footnote to the criticisms of computer-mediated communication made by some print people, among them the writer and critic Sven Birkerts, who, at the very end of his book *The Gutenberg Elegies,* calls on the would-be info-naut to "Refuse it." Refusing it was not really an option for me.

During October, <billg> and I exchanged e-mail two or three times a week. I noticed right away that there seemed to be a peculiar kind of intimacy to talking on e-mail, a sense of being wired into each other's minds, that was not present in telephone conversations (mine, anyway). This was true in spite of the fact (because of the fact?) that our dialogue was as elaborately stylized as a minuet, with no chance for sudden interruption or spontaneous give-and-take. The feeling of intimacy seemed also to come from the shorthand style with which the e-mail was written. It was a form of communication that was neither writing nor speech, but writing used *as* speech, a peculiar in-between kind of expression that I would come to think of as "speak-acting." (The French make a linguistic distinction between a word used as speech, *un mot,* and a word used in writing, *un parole,* but in the English language we don't recognize the difference linguistically, and e-mail seemed to thrive in the gray air between the two.)

After five years of using my computer only as a writing machine, I had grown into the habit of thinking of it as an extension of my own mind, my own private thinking box, and therefore these messages seemed like thoughts that had popped into my head. They were neither "live" like a telephone call, nor three or four days old, like a letter, but seemed to exist inside a new kind of time that moved both slow and fast. I had never thought about the concept of "real time" before. Real time simply *was* time. But now it seemed like there was a new kind of time, which borrowed elements from real time, such as immediacy and spontaneity, but which was controlled by the people using it. E-mail was like a vastly improved version of an answering machine.

During that first week I would sometimes wake up in the middle of the night and lie in bed wondering if I had e-mail from Bill. I had no idea where Gates was when he wrote to me, but it didn't seem to matter. Once, when I was writing him mail on an airplane, I felt physically closer to him than when I was composing from home. Maybe I was thinking of all of the thousands of people who had encountered him on airplanes, restlessly wandering the aisles with his shoes off, or sleeping with a blanket over his head, or staring into the screen of his laptop, writing hundreds of e-mail messages, which would be fired into the network when the plane landed, sending <billg> into hundreds of other computers.

In thinking about what to "say" to Gates I found my mind taking an unexpected, and perhaps not altogether appropriate, spiritual tangent. Perhaps it is true, as the philosopher Jacques Ellul speculated, that the spiritual center of mankind lies close to the capacity for magic, or wonder, that every new technology inspires. Had I been talking to Gates on the telephone, force fields of skepticism and journalistic "independence" would have been activated; ordinary telephonic life would have intervened; and I would have been more inclined to question Gates on, say, Microsoft's alleged alliance with TCI and Time Warner, a scary-sounding combination called "Cablesoft" which was then receiving much comment in the business press (and was later abandoned by the principals). But, bedazzled by the novelty and wonder of e-mail, unconsciously I deactivated force fields, beamed myself up, and spoke as one believer in progress to another whose work seemed to validate that belief.

Subject: How does the future make you feel?

How does the rapid change in the power of microprocessors make you feel? The certainty that microprocessors will grow twice as fast every eighteen months and that nothing in Nature, no fire or earthquake or tidal wave, is powerful enough to stop this from happening. Are you thrilled by this? Do you think that this power is God, as you understand God? Is it possible this power could be bad?

After sending that message I spent a nervous forty-eight hours of checking my e-mail, wondering if Gates thought what I had written was "totally random" (a big put-down at Microsoft; random things did not compute) and had blown me off in disgust. On the second day without a reply, while interviewing a cyber-thinker and computer industry gadfly named Mark Stahlman, who lived in my neighborhood, I asked him whether he thought it would be unhip to send Gates another e-mail before I had received an answer to the previous one.

"I'm new at this stuff," I explained.

Stahlman replied, "Well, hey, you're not a digital guy!"

When I returned home, I logged on and saw that <billg> had responded:

From <billg@microsoft.com>

Feelings are pretty personal. I love coming up with new ideas or seeing in advance what is going to count and then making it happen. I love working with smart people. Our business is very very competitive—one or two false moves and you can fall behind in a way that would wipe you out. Market share does not give you the right to relax. IBM is the best example of this. This is very scary but also makes it very interesting.

The digital revolution is all about facilitation—creating tools to make things easy. When I was a kid I was a lot more curious than I am today—perhaps I have lost less curiosity than the average adult but if I had had the information tools we are building today I would know a lot more and not have given up learning some things.

These tools will be really cool. Say today you want to meet someone with similar interests to talk or take a trip together or whatever? It's hard and somewhat random. Say you want to make sure you pick a good doctor or read a good book? We can make all of these things work so well—it's empowering stuff.

Enough for now.

As I look back at this message now, it occurs to me that Gates did not answer my question. In fact, as I would eventually come to understand, e-mail was a great way of not answering your correspondent's questions, and instead delivering some monologue of your own. But at the time I did not notice this; I pored over Gates's e-mail as if it were a kind of Rosetta Stone—it was the key by which I could decipher the language of the future.

There was no beginning or end to the message. Thoughts seemed to burst from his head *in medias res,* and to end in vapor trails of ellipses. There was no time wasted on words like "Dear" and "Yours," nor were there any fifth-grade composition book standards like "It may have come to your attention that," or "Looking forward to hearing from you." Social niceties were apparently not what <billg> was about. Good spelling and use of the upper case were not what <billg> was about either. I wasn't sure to what extent this style of writing e-mail was idiosyncratic and to what extent it reflected the culture of e-mail, but since Gates was in a position to determine the culture, and I didn't want to sound random, I decided to emulate his style. And here I felt the first of the promises of freedom of speech on-line—the freedom from the conventions of business writing, which are associated in my mind with the letters I used to receive from my father, dictated to and typed by his secretary, sometimes cc-ing my brother at the bottom, signed with the fine-tipped gold Cross pen, black or blue ink, that he always carried in his suit breast pocket: "Dad."

I could not tell whether Gates was impatient or bored with my questions and merely responding to them because this served his interest, or whether he enjoyed talking to me. I sometimes felt this was a game I was playing with <billg> through the computer, or maybe a game I was playing against a computer. What was the right move? What question would get me past the dragon and into the wizard's star chamber where the rich information was stored? I wondered if I was actually communicating with Bill Gates. How hard would it be for an assistant to write this e-mail? Or maybe I was talking to an intelligent agent?

Naaaahhh. . . .

From <73124.1524@compuserve.com>

Subject: Interactive TV as the opium of the people?

Some people are afraid of interactive TV. TV is a drug, goes the argument, and the technology that Microsoft and others are supplying is going to make the drug stronger. People will be inside more than ever, cut off from their neighbors, watching interactive monster truck contests. Or porno. They will pile up large cable and credit card charges. A "T. S. Eliot wasteland . . . of house-bound zombies," as Michael Eisner put it recently in a speech. Do you think this could happen? What difference does it make if you invent smart boxes to deliver dumb programming?

Interactive TV is probably a really bad name for the in-home device connected to the information highway.

Let's say I am sitting at home wondering about some new drug that was prescribed to me. Or wanting to ask a question to my children's teacher. Or curious about my social security status. Or wondering about crime in my neighborhood. Or wanting to exchange information with other people thinking about visiting Tanzania. Or wondering if the new lawn mower I want to buy works well and if it's a good price. Or I want to ask people who read a book what they thought of it before I take my time reading it. In all of these cases being able to reach out and communicate by using a messaging or bulletin board type system lets me do something I could never do before. Assume that the infrastructure and device to do this is easy to use and it was funded by the cable or phone company primarily because I like to watch movies and video-conference with my relatives.

All of the above is about how adults will use the system. Kids will use it in ways we can't even imagine.

The opportunity for people to reach out and share is amazing. This doesn't mean you will spend more time inside! It means you will use your time more effectively and get to do the things you like more than in the past as well as doing new things. If you like to get outside you will find out a lot more about the places that are not crowded and find good companions to go with.

"Good companions to go with"? That sounded a little . . . nerdy. Did we really want someone who sometimes sounded like a ten-year-old boy to be the principal architect of the way people will communicate with each other in the future?

The bottom line is that 2 way communication is a very different beast than 1 way communication. In some ways a phone that has an unbelievable directory, lets you talk or send messages to lots of people, and works with text and pictures is a better analogy than TV. The phone did change the world by making it a smaller place. This will be even more dramatic. There will be some secondary effects that people will worry about but they won't be the same as TV. We are involved in creating a new media but it is not up to us to be the censors or referees of this media—it is up to public policy to make those decisions.

Was this true? It sounded good. But was our e-mail exchange in the true spirit of 2 way communication, or was it just a better way for Gates and his minions to broadcast their agenda to a naive and trusting audience?

I was beginning to understand that you could tell a lot about people from the way they described interactive TV, the information superhighway, the Internet, cyberspace, the matrix, the infobahn, the Microcosm, the End of the World, or whatever you called the THING. What a person thought the THING was, really thought it was, often had as much to do with that person's worldview, memories, expectations from life, and hopes and fears as with the THING itself. Because cyberspace did not really exist in the physical world, it was always necessary to flesh it out with a little of your own imagination. In this way the computer screen was like a mirror. Not a true mirror—more like a mirror that gave you back a vision of the world looking the way you wished it looked. Every user's desire to live in the pluperfect, the would have, was part of what cyberspace was all about. For Newt Gingrich the THING was a free, market-driven, libertarian, unregulated, decentralized, grass-roots democracy, in which the federal government was completely uninvolved. For Vice President Albert Gore, on the other hand, the THING was a government-assisted enterprise in the spirit of the great social programs of the New Deal, but without the public spending. Gore's father, Senator Albert Gore Sr., who like Al Jr. had been a Tennessee senator, had been one of the driving forces behind the creation of the interstate highway system, in the 1950s, and Al Jr. seemed to be using the information highway metaphor not only to conceptualize computer networks but also to model his relationship with his father. When James Billington, the Librarian of Congress, was asked how he envisioned the THING, he said, "My vision of the future is my earliest vision of a big library, when I sat in the Slavic reading room of the New York Public Library between Trotskyites and monarchists, Stalinists and misplaced liberals. Had they been out on the street they would have been shooting each other, and here they were reading together." A classic case of the would have.

My notion of the THING was that it was a giant electronic brain. When I was seven or eight I used to think that there should be some giant brain that could answer questions you really wanted to know, like what the deepest snowfall *ever* was. Since such an answer *could* be known, it *should* be known, and it seemed like a definite flaw in the universe that there wasn't such a brain. That was what the THING meant to me, when I first encountered it. It was a way of being able to find what you needed and to make good use of it. I imagined coming into a summer house that had been shut up all winter, and finding in the cupboard cornmeal, canned olives, tomato paste, and canned clams, along with a few other durable staples, plugging my laptop into the phone line, entering the ingredients into the THING, and coming up with the perfect meal.

About ten days into my correspondence with <billg>, I began to feel the inevitable fade that signals the start of the second phase of the typical e-mail relationship, when your excitement over the possibilities of this marvelous new toy—you can say anything you want to each other!—settles into the realization that you don't have that much to say. In short, I started talking about sports:

Subject: sports
I heard you had taken up playing golf. Is that true? Is it mainly for business, or recreation? Do you play any other sports? Tennis? Do you waterski on Lake Washington?

I don't really play golf. I played once a few years ago in a scramble tournament and did reasonably well and my whole family plays but I haven't taken the time to get very good. My handicap would be around 20 or so. Who knows, I may get serious about this game at some point.

I play tennis quite a bit. I waterski. I snow ski. I used to do all of these more than I do now. I have a great waterski boat and have ended up teaching a lot of people how to waterski which is fun. I broke my leg once waterskiing when I was young just because I was showing off on a race course. Lake Washington where I live is great for skiing in the morning. Hood Canal where my family has gone every summer and I have a summer place is even better. My best waterskiing recently was in the Indian Ocean off on Nemba island during my Africa trip.

I have a pool that I swim in and my house is on the lake. My girlfriend and I just got new kayaks but we are just learning—it's fun on Lake Washington.

My responses are slower now because I am off at Hood Canal on a 'think week' where I go away and read and try out products and walk around and think and hopefully write some memos about ideas I have and strategies. I do this twice a year to really catch up on things and think ahead a little.

After I had sent a message, the urge to check for a response would soon be stabbing me in the gut. Later, when friends of mine who were new to e-mail would send their first message to me (because mine was the only e-mail address *they* knew), they would often telephone to see if I had received it—to see if it had "worked." In a sense it had not worked, which was why they were calling. The immediate gratification of hearing the other person's voice that you got in a telephone conversation, in which the communication and the response were virtually simultaneous, was missing in e-mail; you might not even realize this was what was missing, but that was one reason why you kept checking your e-mail. You craved a response.

Checking my e-mail soon became a way of dividing time in the day, the way smoking cigarettes used to be. I'd check it first thing in the morning, at lunchtime, once during the afternoon, once after dinner, and one last time before going to sleep—just to see what the electronic cat had dragged in. A new window of possibility had opened inside my everyday life. Sometimes, as I was checking my mail, Stahlman's phrase, "Well, hey, you're not a digital guy!," would pop back into my head, and I would stop and ask myself, Was I thinking like a digital guy? What is a digital guy? Is a digital guy who I want to be? Are digital guys what nerds will molt into when the information highway reaches everyone's door? One night I was at home listening to music, doing this geeky dance I like to do, and wondering whether the Wall Street types across the street were watching me and thinking about what a weirdo I was, when I thought, Why should I care about those guys? I took this as a sign that, for better or worse, I was becoming a digital guy.

From: <73124.1524@compuserve.com>

You love to compete, right? Is that where your energy comes from—love of the game? I wonder how it feels to win on your level. How much do you fear losing? How about immortality—being remembered for a thousand years after you're dead—does that excite you? How strong is your desire to improve people's lives (by providing them with better tools for thinking and communicating)? Some driven people are trying to heal a wound or to recover a loss. Is that the case with you?

From: <billg@microsoft.com>

It's easy to understand why I think I have the best job around because of day-to-day enjoyment rather than some grand long-term deep psychological explanation. It's a lot of fun to work with very smart people in a competitive environment. Microsoft gets to make its products a lot better every 18 months. We get to hire the best people coming out of school and give them challenging jobs. We get to try and figure out how to sell software in every part of the world. Sometimes our ideas work very well and sometimes they work very poorly. As long as we stay in the feedback loop and keep trying it's a lot of fun.

It is pretty cool that the products we work on empower individuals and make their jobs more interesting. It helps a lot in inventing new software ideas that I will be one of the users of the software so I can model what's important. Its fun to meet with users and see how they use our products in different ways than we expected. It's great to see how clever people are when they are given an excellent tool.

Just thinking of things as winning is a terrible approach. Success comes from focusing in on what you really like and are good at—not challenging every random thing. My original vision of a personal computer on every desk and every home will take more than 15 years to achieve so there will have been more than 30 years since I first got excited about that goal. My work is not like sports where you actually win a game and it's over after a short period of time.

Besides a lot of luck, a high energy level and perhaps some IQ I think having an ability to deal with things at a very detailed level and a very broad level and synthesize between them is probably the thing that helps me the most. This allows someone to take deep technical understanding and figure out a business strategy that fits together with it.

It's ridiculous to consider how things will be remembered after you are dead. The pioneers of personal computers including Jobs, Kapor, Lampson, Roberts, Kaye, are all great people but I don't think any of us will merit an entry in a history book.

I don't remember being wounded or losing something big so I don't think that is driving me. I have wonderful parents and great siblings. I live in the same neigborhood I grew up in (although I will be moving across the lake when my new house is done). I can't remember any major disappointments. I did figure out at one point that if I pursued pure mathematics it would be hard to make a major contribution and there were a few girls who turned me down when I asked them out.

From <71324.1524@compuserve.com>

I heard you read a lot.

From: <billg@microsoft.com>

I like biographies and autobiographies. I like English literature and history. Certain writers I decide to read everything from:

Heinlein, Fitzgerald, Hemingway, John Irving, Salinger, Fowles. . . . I love history of science—*The Second Creation, The Eighth Day of Creation.* . . . I like science—*Molecular Biology of the Gene* by Watson et al. 4th edition, Dawkins, Gould. . . . Sometimes I get into a person or a period—the making of the atom bomb, Feynman, Da Vinci, Napoleon, Gauss, Von Neumann, Turing, Roosevelt, Wilson, Ford, Kennedy. I even read biographies of people just out of interest—Getty, Paley, Hughes, Neurath, Onassis, even Madonna.

In terms of history my knowledge is spotty—pretty good before 20k BC (since it's science), reasonable on Greece, OK on Rome, but mostly post 1780. Starting with the French Revolution until Napoleon died is a great period. Victorian England was a great period—lots of great novelists and thinkers—Ricardo, Malthus, Marx, Dickens, Thurber. . . . I have read a lot on U.S. history— still curious to understand economic thinking over the ages. . . . Some puzzles intrigue me, like how much did the U.S. know about what was going on in Germany in the late 30s? How did we mess up so badly in Vietnam? Could the British have maintained more of their leadership role?

Fitzgerald is a real favorite. His life story avoids his being any kind of role model but he was a great writer. *Tender Is the Night* might be his best but the story of Gatsby is very appealing. Gatsby had a dream and he pursued it not even really thinking he might fail or worse that what he dreamed of wasn't real. The green light is a symbol of his optimism—he had come so far he could hardly fail to grasp it. I don't think I ever spoke about this in an interview but maybe I did. At the end Fitz is reinforcing what a romantic figure Gatsby is. It's also sort of about America but I think of it more in terms of the people.

At the end of one message, I could not restrain myself from writing:

This reporting via e-mail is really fascinating and I think you are going to come across in an attractive way, in case you weren't sure of that.

Why did I write that? I certainly wouldn't have said that to a businessman I was interviewing on the phone. It rushed out of me in a gust of sympathy, perhaps as much a feeling for the technology that united us as for <billg>. But I did like <billg>. I had begun to look to him as my digital helper. The voice of <billg> was curious, callow, beguilingly boyish, and full of intimate shorthand. Although he was light-years ahead of me, technically speaking, he seemed to be feeling his way along in this new world, just as I was. It was hard for me to believe that this little critter I had in my computer could be the spirit of world domination. He seemed so harmless, not at all like the Big Brother figure some people feared he could be. (Of course, if Gates were Big Brother, this would be precisely how he'd want to appear to people: not big at all, but as quiet little <billg>.) When I imagined <billg> I thought of that picture of the thirteen-year-old Gates in the computer room at Lakeside. And because I felt I knew who that boy was, the voice of <billg> touched me.

From: <billg@microsoft.com>

I comb my hair every time before I send e-mail hoping to appear attractive. I try and use punctuation in a friendly way also. I send :) and never :(.

II. **The Bear** ◊

Toward the end of October, my hour in Bill Gates's physical presence arrived. As we shook hands I looked for signs of <billg> in his pale, freckled face, which did look oddly boyish, though not because Gates looked younger than his age. In some ways he looked older than his age, like a very old little boy. He said, "Hello, I'm Bill Gates," and emitted a low, vaguely embarrassed-sounding chuckle. I wondered if this was the sound one e-mailer makes to another when they finally meet in real space. The e-mail we had exchanged was part of the moment, but at the same time our correspondence seemed to exist in a world too remote from this one to acknowledge it now. Yet the thought of what had already transpired between us gave meaning to the polite formalities and conventional sniffings around each other which we engaged in as we settled into the interview mode.

Gates's office was exactly twice as large as the offices of junior employees. His carpeting was a little richer than the carpeting in other offices; otherwise there was nothing fancy about the place. A large monitor sat on his desk, and on the walls behind were pictures from important moments in Gates's career, many of which coincided with important moments in the history of the personal computer. There were also pictures of Bill's two sisters and of his mother and father. There was very little paper around.

Gates wore a green shirt with purple stripes on it, brown pants, and black loafers. As we stood chatting for a while he pushed his pelvis forward slightly, wrapped one arm around his body, the other arm occasionally going up into the air as he talked, kind of flying up, with the palm outstretched, then settling again somewhere on his chest. His heavy lips contorted into odd shapes as he talked. His voice was toneless, with a somewhat weary note of enthusiasm permanently etched into it, and his vocabulary was bland—stuff was cool, neat, crummy, super, supercool.

That morning in the hotel I had put on a necktie, which I hate to do, thinking that one ought not go with an open collar to one's first meeting with the world's richest man, but in the Microsoft parking lot the voice of <billg> entered my head and I stripped my tie off and stuffed it into my pocket.

We sat down in the meeting area in the front part of Gates's office, at right angles to each other. He did not look at me very often: he either looked down as he was talking, or he lifted his eyes over me to look out the window at the campus. But the e-mail we had exchanged felt like a form of eye contact. The angle of the light caused the purple stripes in Gates's shirt to reflect in his glasses, which in turn threw an indigo tinge into the dark circles around his eyes.

I asked, "Do you worry that your wealth is going to corrupt you?"

"Absolutely." Gates sat upright and raised his arms in the air. "Absolutely. Hey. Being in the spotlight is a corrupting thing. Being successful is a corrupting thing. Having lots of money is a corrupting thing. These are very dangerous things, to be guarded against carefully. And I think that's very, very hard to do."

"How do you do it?"

"I'm very close to my family. And that's important to me. It's a very centering thing. We get together as a family a lot. Melinda wants when we have kids to have a normal environment for them. So we'll mutually brainstorm about how to do the best we can at that." Gates thought for a while, then said, "I am a person who is very conscious of, like, why don't I have a TV in my house? I think TV is great. When I'm in a hotel room, I sit there and try all these new channels and see what's going on. I probably stay up too late watching stuff. TV is neat. I don't have a TV at home, because I would probably watch it, and I prefer to spend that time thinking— or, mostly, reading. So I'm pretty conscious of not letting myself get used to certain things."

While Gates was talking he rocked his upper body down toward his thighs, rocked back up, and rocked down again.

I had been wondering to what extent Gates's interest in technology was like mine, an attempt to disprove evidence of the possible meaninglessness of the universe by latching onto material progress with special vehemence; but I didn't want to sound too random in asking about this. So I asked:

"Do you consider yourself a puritanical person?"

"Oh, no no no. I'm not a Puritan. Hey, if I were a Puritan—" He grinned, apparently mentally flipping through a sequence of acts he had committed that would have shocked Jonathan Edwards. "Okay, it's a little bit like this," he said. "I go to a baseball game, and I'm having a good time, watching the game, but then I feel myself getting drawn in. I start wondering, Who are these guys? Who are the good ones? How much are they paid? How are the other teams compared to this one? How have the rules changed? How do these guys compare to the guys twenty years ago? It just gets so interesting. I know if I let myself go to ten games I'd be addicted, and I'd want to go to more, and there's only so much time in the day. And, frankly, it's easy for me to get interested in anything. I think, Gosh, am I going to get good at tennis? Well, we got these kayaks recently. I think, you know, Are we going to get into that? I was just in Africa. I think, Should I do my next two trips there—there's just so much there—but I'd sort of like to go to China, and actually I think I'll end up doing that for my next big trip, in two or three years. So there's all these choices, but time is this very scarce resource."

So far, the conversation had been pleasant and interesting. Things seemed to be going well. I was getting some psychic e-mail from my father, and also from his father, which said, <Good Boy>. But at the same time I wanted to see the bear. In articles I had read, people had spoken of Gates in meetings, purpled-faced, spraying saliva into the face of some hapless employee as he screams, "This stuff isn't hard! I could do this stuff in a weekend!" Or "That's the stupidest fucking thing I've ever heard!" In preparing for the interview I had thought up some questions that I thought might anger the bear. Finally one of Gates's monologues slackened and his eyes went up to mine, and I knew I had my chance.

Around the time <billg> and I were e-mailing each other, the federal government was in the process of bringing a series of antitrust actions against Microsoft. At the heart of the government's case was the question of whether Microsoft's dominance of the PC operating system market constituted an anticompetitive advantage in the software industry. Microsoft's critics, including many of its competitors in the software industry, argued that Microsoft's control of the standard operating system gave it an advantage over rival makers of applications because the Microsoft application designers had special insider knowledge of the Windows source code, knowledge that they could use to make their products perform better than the products of their competitors. Some competitors even charged that Microsoft secretly embedded code in its operating system software which caused other, non-Microsoft products to crash the machine—charges Microsoft hotly denied.

"Some people say that the government wouldn't have brought the antitrust action if they didn't already have a good idea that there were antitrust violations," I said, looking at Gates. Suddenly all the soft planes in his face contorted into an expression of pure sarcasm.

"I think you're a little *confused,*" he sneered. "You're saying, before they read even a single piece of paper, they judge what kind of case they have?" He seemed to be slightly choking on his disgust for my stupidity. "I think you're confused," he said again. Now a mocking pedagogical tone came into his voice. "The Justice Department *chose* to get the information to *decide* what to do. Saying they have a pretty good case before they've read anything—is that how these things work?"

Abruptly the emotional boundaries of our encounter seemed to have been greatly expanded by the e-mail that preceded it. I remembered a line from Gates's very first message: "E-mail is not a good way to get angry at someone, because you can't interact." Maybe this was the way lots of people would communicate in the future: meet on-line, exchange messages, get to know the lining of each other's minds, then meet face-to-face. In each other's physical presence people would be able to dispense with a lot of the polite formalities that clutter their encounters with one another, and say what they really meant. It would be like seeing a live performance by a band whose recordings you knew well. Electronic communication wouldn't reduce face-to-face communication; it would focus it.

In encounters with grizzly bears, you were supposed to stand your ground. If you flee the bear will think you're prey and pursue you, and you can't outrun a bear. Going by the book, I said, "Someone at the Federal Trade Commission could have *told* someone in the Justice Department that the case against Microsoft was strong." But this seemed to make the situation worse. Gates began to rock furiously. "Look," he said, not looking at me. "The Department of Justice is looking at these files. You know? It's *justice?* You're supposed to have *facts* before you decide things?"

I looked down and away, which is another thing I have read you are supposed to do in the event of a bear attack, and I changed the subject.

As we were saying good-bye, Gates said, pleasantly, "Well, you're welcome to keep sending me mail."

I walked out to my car, drove off the Microsoft campus, and headed back to Seattle. When I got back to my hotel (where I had been put in "the Dale Carnegie Suite"), I logged on and saw I had e-mail from Bill. He was responding to e-mail I had sent before our meeting, asking what he thought of Henry Ford. The message had been sent about two hours after I left his office. There was no reference to our having just met. Gone was all trace of the bear; the quiet, musing voice of <billg> had returned.

Ford is not that admirable—he did great things but he was very very narrow-minded and was willing to use brute force power too much. His relationship with his family is tragic. His model of the world was plain wrong in a lot of ways. He decided he knew everything he needed to fairly early in his life.

Chapter Three   The Great Migration

I. **E-mail from Mom** ◊

Elizabeth Toomey Seabrook was always a good letter writer. (Her gift for writing in general won her first prize in the senior essay writing contest at Columbia College in Missouri, and later carried her through a successful career as a newspaper writer, which culminated in a column she wrote for the United Press wire service in the 1950s, before she married my father and gave up her career.) I relied on her letters when I first went away to boarding school, and earlier, when I spent eight unhappy weeks at Camp Wild Goose, in Maine. Her letters were clear, funny, intelligent, chatty—a readable version of her voice—and they came in the mail twice a week, once more than I was allowed to make telephone calls home, and the letters were less sad than the calls, because they didn't end with that awful *click.*

Although I had stopped writing letters to her when I was in college, my mother had continued to write to me, though less often;

now her correspondence tended to consist of hastily scrawled notes containing her thoughts for story ideas for me, sometimes attached to a clipping from the local newspaper. "I think you should write a story about *balls*. Baseballs, softballs, golf balls, tennis balls. . . . Balls strike me as very interesting, somehow." By 1993 these hand-addressed envelopes were the only personal mail I could look forward to. No one else wrote me letters anymore. Going to the mailbox had ceased to be a moment of adventure in the day and had become instead a chore of catalogue and bill hauling. If we went away two or three days, I could feel the weight of the junk mail pushing against the mailbox door as I inserted the key, and, if I wasn't prepared for it, the coiled cylinder of junk would spring out, raining pulp and gloss at my feet.

That fall, when my father asked me if I had any ideas for what to give Mom for their thirty-seventh wedding anniversary, I suggested a PowerBook. My mother, seventy-two at the time, had never used a computer of any kind before, but I had confidence in the intuitive genius of the Apple software designers. I went down to J&R Computer World again and got the model that I had been lusting for recently, which had a better screen than mine. I installed a Global Village modem myself, which gave me a hardware thrill I had not had before with a computer, loaded the CompuServe software, and set Mom up with a CompuServe account. Finally, I recorded on the hard disk Dad's voice (by plugging a microphone into my PowerBook and holding the speaker next to the earpiece of the telephone) saying "Happy Anniversary!" instead of the "Wild Eep" and other factory-installed sounds that the computer makes when it wants to get your attention.

My father presented the PowerBook to Mom at our place. I had put the computer in Sleep mode, and as she opened the wrapping I touched a key and it woke up.

"Happy Anniversary!" the machine said.

"Oh, mercy, it's a talking computer!" my mother said. She looked at my father. "Thank you, darling, that's very sweet of you, but I'll never be able to use it."

"It's really pretty easy, Mom," I said. "I'm going to teach you." I took her through the basic operating procedure, and showed her how to run the animated Macintosh tutorial which I remembered from my novice days six years before. Then I explained about e-mail.

"Okay, the main concept to grasp here is that the computer is not only a typewriter, it's also a telephone. Okay? Repeat that."

"It's not only a typewriter, it's a phone."

"Right. Except you don't actually talk to other people with it, you write them letters and they write you back."

"How do I do that?"

"You write the letter on the computer, just like you'd write it on a typewriter. Then you plug the phone line into this hole here in the back of the computer. The letters come in and go out through the phone wire. So, if you want the letter I wrote you—"

"You wrote me a letter? I didn't get it."

"That's because I didn't put it in the mail. I sent it to your e-mail address."

"You haven't written me a letter in years!"

"Well, I have now."

"And you sent it through the phone line?"

"Right."

"Like a fax," my father said, hopefully.

"Right, like a fax, except the mail doesn't come to your computer directly, like it would come to your fax machine. To get my letter you have to plug your computer into the phone line and dial up the number of the computer where your mailbox is. See, that way you can get my letter wherever you are in the world, whenever you want, which is better than a fax."

I showed my mother how to open her CompuServe application and how to log on.

"Okay, see that little number under the envelope? That means you have one new letter. So, okay, now put your hand on the trackball."

"The trackball? Oh, yes, the trackball. Okay, got it."

"Now point the cursor at the envelope and double-click on it. No, double-click."

My e-mail to her appeared.

"Voilà! Magic!" I said.

But my mother wasn't interested in the technical miracle of how my letter got into her computer; instead, all her attention was on the news in the e-mail itself.

"What do you mean you may not be able to come for Thanksgiving?"

"Well, it's not definite but I have so much work to do that—"

"You have to come. My heart will be broken if you don't."

"Okay, okay, we can talk about this later."

It took my mother about two hours to learn to write and send me a message that said, If you don't come to Thanksgiving my heart will be broken.

And that was only the beginning.

From: <72733.3174@compuserve.com>
_____

Dear John,

I don't suppose you would cut your hair? You are much more handsome with short hair as it gives your head a better proportion to the rest of you when there is a gently curling couple inches of hair on top. I've got pictures to prove it!

From: <72733.3174@compuserve.com>
_____

Dear John,

Have you re-thought going to Vermont and sleeping out since

the weather has turned so awful again? You might wake up frozen
to death.

When I didn't respond to an e-mail of hers for a while, I got:

From: <72733.3174@compuserve.com>

I thought the whole idea was you were supposed to answer
e-mail. You were the one who told me that.

My God, I thought, what have I done?

But Mom grokked it. She knew right away how to use the machine to throw her voice. It seemed like something I had lost had
been restored to me—her voice, in writing. My mother was smart,
American smart, not exactly Microsoft smart, not necessarily efficient, but humorous and honest and independent and full of common
sense, and all of those qualities came through clearly in her e-mail.
Unlike my father, she was a westerner, from Spearfish, South Dakota; her grandfather, Daniel Joseph Toomey, had homesteaded a
place in the Black Hills in the 1870s. Her voice had the sunny,
outward-looking spaces of the West in it, different from the colder,
inward-turned eastern spaces of my father's side of the family. I got
from her the liberal ideas my father enjoyed disputing. When I was
a boy she had decided for me that I needed a hero, and that my hero
was going to be Thomas Jefferson; she had taken me down to
Monticello, where she bought me the little copper bust of Jefferson
that sat in my room all during my boyhood. I'm pretty sure Jefferson
would have been proud of her.

No one else I subsequently met on the Net could touch the woman
for carefulness of thought, grammar, spelling, punctuation, and
timeliness of response. I saved all her e-mail, and hope that this
floppy disk version of her will survive long enough for her grand-
children and great-grandchildren to appreciate how much energy
she devoted to agitating for their existence.

I was assuming my mother saved my e-mail, too, until one day on the phone when I asked her about a message I had sent her, and she said, "Oh, I trashed that."

"What? You trashed my e-mail? I thought you saved all my letters!"

"Oh, well, I save your letters, sure, but somehow e-mail doesn't really seem worth saving."

II. Is this communicating or what? ◊ ◊ ◊ ◊ ◊ ◊ ◊ ◊ ◊ ◊ ◊ ◊ ◊

My article about Bill Gates was published in the *New Yorker* shortly after New Year's Day, 1994. It included some of our e-mail to each other, as well as both our e-mail addresses. Gates had told me in his first e-mail that his address was known "very broadly," and I had had no trouble getting it myself, so I figured that publishing his address would be no big deal. But it turned out that in doing so I was committing an act of info-terrorism. (Later, when the article was reprinted in Japanese *Playboy,* the editors made Gates's e-mail address a big juicy-looking double-page spread in the magazine, the Information Age equivalent of a centerfold.) Within a few weeks, Gates had 5,000 messages waiting in his mailbox, overwhelming even the capacity of the he-man of e-mail to process them.

To: <billg@microsoft.com>

From: <TECHNOWEENIE@AppleLink.Apple.COM>

Dear Mr. Gates,

My name is Tom Williams. I am CEO of Desert Island Software (DIS), a CD-ROM edutainment company (developer/publisher), based in Victoria B.C., Canada.

I admire the way you started and when I heard about your days at Lakeside, I realized that I am much like you when you were about my age (I am 14).

I am about to incorporate my company and I am writing my business plan. I am also finding people to sit on the advisory and directors board of Desert Island Software. It would be an honor to have you join this advisory team. . . .

To: <billg@microsoft.com>
From: <mstrright@aol.com>

Bill,

I'm a 35-year old businessman and screenwriter, and I am a genius.

I have a moderate background in information technology from my college days, but I found I had too much talent working with people to spend my days writing code.

Since leaving NYU I've built up several businesses and am now wealthy.

But I'm not really good at anything, not in a world-class way, anyway . . . not in the way I should be.

And I should be, because I'm incredibly talented.

To: <billg@microsoft.com>
From: <curth67100@aol.com>

Read the article concerning you in *New Yorker*. Re your favorite books: recommend that you read *The Waves* by Virginia Woolf. Narrative consists of 6 "disembodied" voices—a lot like e-mail. Might be of interest.

To: <billg@microsoft.com>
From: <jaklasse@sciborg.uwaterloo.ca>

Mr. Gates,

I may be coming to Seattle next summer to work for a nonprofit environmental group called "Friends of the Earth"—can you suggest a good restaurant I might try while I'm in Seattle.

I also published my own e-mail address in the article. On the morning the article was published, I checked my mailbox and found it stuffed with twenty-nine messages. In the next two weeks I received 396 electronic messages from readers, almost all of them from strangers. Over the same period I received eight phone calls about the article, seven letters, and one fax. The calls were mostly from writer friends, who didn't have the time or the energy to write a letter, and didn't have e-mail. The only telephone call from a stranger was from Bill Gates's tailor, who objected to my characterization of Gates as a sloppy dresser, and wanted me to know that Gates's khakis were handmade, his shirts were Sea Island cotton, and that he, the tailor, had personally made five beautiful suits for Gates, not counting a double-breasted tuxedo with a shawl collar that Bill had worn recently at a Seattle reception for him and his new wife, Melinda, at which Natalie Cole performed.

From: Don Hinkle, <73776.2504@compuserve.com>

Just read your Gates profile in *NYORKER*. A couple of questions: How many drafts did you do? Which draft did you submit? Did you need to do much revision after submission?

From: <oconnort@acf2.nyu.edu>

When a person like John Seabrook can pop a message to "billg," what does that do to Machiavelli's model of power and of access to the person who holds the power? If tomorrow's corporate leaders choose to be insulated (as Gates apparently is not), will that choice help them preserve their power? Or will it be a strategic disadvantage? What really happens when the Prince gets wired?

From: Steven Schlosstein, <76550.1416@compuserve.com>

The digital revolution is the obverse of what Marshall McLuhan proclaimed 30 years ago. We are nowhere near becoming a global village; in fact, quite the contrary, though the medium is still the message. The digital medium fragments the world, making decentralization one of the most powerful forces driving the information economy. The collapse of the Berlin Wall and the implosion of Soviet and Eastern European command economies had nothing to do with Reagan/Bush and everything to do with the decentralizing power of the chip.

From: <71331.3662@compuserve.com>

The simple truth of the matter is that Bill Gates wields an amount of power and wealth that would cause Citizen Kane to blush, yet he wields it in very much the same kind of isolation that Orson Welles did in the movie. That kind of isolation and wealth is a very dangerous psychological culture dish for the formation of ideas that have such a far-reaching resonance in the cyberworld that you have written about.

Many people who are excited about the role of computers in the expansion of human consciousness are afraid of that power. To them it is a black and reactionary force. You saw Bill Gates as some sort of oracular Yoda perched on a misty summit of cyber wisdom to whom even a lowly journalist can gain access via the miracle of cyberspace—a miracle that Gates implicitly had a large hand in creating. But many people feel that the reality of Bill Gates is diametrically the opposite of that image. That is, someone who understands that digital interconnection between humans can be an unlimited source of power and wealth for the person who understands how to control its access. Objectively examined, the career of Bill Gates has been a nearly perfect study of one man's attempt to consolidate that control.

From: <71240.524@compuserve.com>

Hi! I just finished reading your article in the *New Yorker* about Bill Gates and e-mail, soooo I decided to write you a letter and see what happens.
Write Back Please!
Melinda Gottesman
14 yrs. old

From: <peter911sc@aol.com>

Real problem with the Information Superhighway is typified by this letter: God only knows how many idiots like me will tie up your time with responses. Will there be unlisted numbers on e-mail? Will we be able to block mail from pests . . . or junk e-mail from those who send sequentially? It doesn't cost ANY-THING to e-mail—YET—so junk mailers, telemarketers will have a field day.

Or, let's say for a moment I'm not an idiot. Just "curious," like the young billg. Read an article, question pops into mind . . . BINGO . . . look how easy it is to either seek an answer or bother someone else with my maunderings. In fact, your article impelled me twice to the computer in the living room . . . once to congratulate billg on his marriage—his e-mail address was not known to me previously, and you made it sooo easy. I also e-mailed the address to a similarly spontaneous friend across the country, so he'll probably send bill a "card" too . . .

Multiply this response by x-percent of the *New Yorker*'s reader-ship and you are responsible for the wasting of x times y time. And the POINT IS: We've both wasted all this time! Was it as good for you as it was for me? (Or, as I tell my wife, at least she knows where I am and no illegal substances are involved.)

E-mail will be like cable TV one of these days. Too many channels and nothing's on.

Lots of people wrote me e-mail to ask how much e-mail I was getting.

From: Carol McAlpine, <76645.1453@compuserve.com>

I'm still trying to figure out how, precisely, e-mail is different from other communications, why I don't mind writing e-mail but hate writing letters, and why e-mail makes me feel as if I am free to send a bunch of nonsense to you, a person I've never met & likely will never meet, when I would never do so if it meant writing it down on paper & putting it in the snail mail.

From: <70733.3625@compuserve.com>

E-mailing has revived the art of letter writing, and I love it, but in some ways it's _too_ easy. I suppose I'm old-fashioned and like some hierarchy, some distance. I remember writing a letter to Pauline Kael and she wrote back; it was wonderful, but part of that was my sense that here was this person who I had admired all my life and thought I'd never touch in any way and she's writing a letter to _me_. But . . . this is not at all the kind of interaction you have with email. It just isn't the same.

From: <sitter1@husc.harvard.edu>

Is this communicating or what?

If a message was complimentary, I clicked the "Forward" icon and sent it on to my mother.

I just finished reading all the messages you sent me. I nearly cried several times I was so thrilled at the reaction and the compliments to you. And laughed at some of them. How come the spacing was so weird with only one word on some lines? I trust you will copy them all and take the stack in to Tina.

III. **The Age of E-mail** ◊

Part of the beauty of the Internet, which was celebrating its twenty-fifth birthday in 1994, was that its inventors had not anticipated that ordinary people would use it to talk to one another. The Internet was not the product of one individual inventor's genius, which was the model of invention in the nineteenth century, nor was it a coordinated team effort, like the creation of the microprocessor, the quintessential twentieth-century invention. In some respects the Net was more like the fossiled backbone and vertebrae of some extinct leviathan, except that instead of bone it was composed of glass, silicon, copper, and plastic, and inside its body cavity new life forms were teeming.

In the late 1950s and early 1960s, a tide pool containing the basic components of the Internet collected in graduate schools and military research centers around the United States. The bolt of lightning that struck the soup and created the Net was the Cold War. Two main branches of computer science converged in the Net: one, the information theory and data flow technology that gave computers a way of talking to one another, and two, the design of the network itself. The work for the first part was based on Claude Shannon's seminal paper "A Mathematical Theory of Communication," published in 1948, which established the discipline of information theory—the science of sending data in the form of ones and zeros, or bits. In the early 1960s, "demand multiplexing" was invented by Leonard Kleinrock at MIT. In a multiplex network, data leaves your computer and enters another computer known as a router, which breaks the data up into packets, each of which contains the address of the computer you want the data to reach. In a telephone conversation, which took place on what is known as a "circuit-switched network," a small piece of the network was set aside for the exclusive use of the parties having a conversation, until they chose to give it up. This made for a clear conversation, but it was an inefficient use of the resources of the network. Data transmissions didn't need to be as clear as voice conversations, and they were bursty, unlike telephone conversations, so it made more sense for that part of the network to be available to more than one user at a time. Demand multiplexing was ideal for this, because many different packets could use one part of the network at once. You got to use a piece of the network only when you needed it (i.e., when you had something to send); when you momentarily stopped needing it, someone else could grab it and send data during your silence.

The second part of the Internet, the work on the design of the network itself, was done at the Rand Corporation, in Santa Monica, California, the leading Cold War think tank. In the early sixties the Defense Department asked Rand to come up with a design for a communications network that could survive a nuclear holocaust. The telephone network was the wrong model, because the switches and controls were centralized. For example, if a terrorist blew up the big AT&T regional switching headquarters, a thirty-story windowless dark-brown slab that was a prominent feature of the landscape of downtown Manhattan, which I could see by turning my head away from my screen and looking out of my real "window," he would be able to take down a large part of the Eastern Seaboard's telephone network. The disaster that struck AT&T's long-distance network in 1990, which knocked out the intelligent 800 service nationwide for a day, was caused by a single flawed line of software code running through a computer in that building.

Paul Baran, a Rand fellow, proposed solving this problem by designing a distributed, many-to-many network, in which the responsibility for relaying information was spread equally among all the nodes on the network. If the most direct route between the computer that was sending the information and the computer that was receiving it was occupied or blasted away, the message would keep flying back and forth among other nodes on the network until it found a way around the blockage and reached its destination. Baran called his idea "hot potato routing."

In 1968 a Boston electronics firm called Bolt Beranek and Newman was contracted to build the first router. When it was finished it was flown across the country and hooked up to another computer in a lab at UCLA supervised by Kleinrock, on Labor Day weekend, 1969. The big moment had arrived: it was the first time one computer had tried to speak to another computer. Kleinrock described this moment to me in e-mail, sent via the offspring of that moment, twenty-five years later:

From: <lk@CS.ucla.edu>

What did it mean to be the FIRST node on a NET? Well, it meant that a piece of equipment was wheeled into my laboratory, this equipment (the size of a telephone booth) had the job of doing all the network switching, etc., and was called an Interface Message Processor (IMP). The trick was to get my time-shared computer (referred to as a HOST computer) to talk to this beast, the IMP. To make a long story short, after a furious few months preparing for the arrival of the IMP, upon its arrival, we successfully transmitted the first bits back and forth from our HOST to the IMP. There was no message, per se, just bits and bytes. But that was the "birth of the Internet," its first gurglings, if you will.

The second critical event occured about a month later, in October, 1969, when a second bona fide HOST was attached to its own IMP at Stanford Research Institute, and a long-distance telephone line connected UCLA and SRI. As soon as SRI attached to its IMP, under my directions, one of my programmers, Charley Kline, arranged to send the first computer-to-computer message. The setup was simple: he and a programmer at SRI were connected via an ordinary telephone line and they both wore headsets so they could talk to each other as they observed what the real network was doing. Charley then proceeded to "login" to the remote SRI HOST from our UCLA HOST. To do so, he had to literally type in the word "login"; in fact, the HOSTS were smart enough to know that once he had typed in "log," then the HOST would "expand" out the rest of the word and add the letters "in" to it. So Charley began. He typed in an "l," and over the voice headset, told the SRI programmer he had typed it in (Charley actually got an "echo" of the letter "l," from the other end and the programmer said "I got the l." Then Charley continued with the "o," got the echo and a verbal acknowledgment from the programmer that it had been received. Then Charley typed in the "g," and told him he had now typed in the "g." At this point, the SRI machine crashed!!

Some beginning!

One small keystroke for man, one giant carriage return for mankind.

The Defense Department's Advanced Research Projects Agency (ARPA) began linking supercomputers at different points in the country together, using leased long-distance telephone lines. The number of nodes on the Net grew from four at the end of 1969 to thirty-four in September 1972, sixty-two in 1974, 111 in 1977, and 213 in 1979. In 1974 a student of Kleinrock's named Vinton Cerf, working jointly with Robert E. Kuhn, developed a protocol known as TCP that made it possible for every computer on a network to talk to every other computer on any other network. Although the original intention of the ARPAnet's designers had been to give remote computer users access to supercomputers—to allow an aeronautical engineer in Seattle, say, to do flight simulation tests on a more powerful computer located in Chicago—it turned out that Moore's Law, by distributing computing power more evenly across the whole network, soon made resource sharing less important. Information sharing, on the other hand, became more important. Before long, packets of "data" containing engineers' political opinions and favorite science fiction books were being added to the bitstream carrying their research back and forth.

As early as 1973, the bulk of the traffic on the ARPAnet was network e-mail. For the computer researchers and technicians who had access to the Net, e-mail (as it came to be known) was like a free long-distance phone call. When Len Kleinrock was in England, speaking at a conference at the University of Sussex in 1973, he left his razor in the dormitory where he was staying, and sent a message over the ARPAnet asking for it back. This moment was perhaps the real birth of the Internet.

Just as the telephone was one of the first successful inventions to make use of electrical circuits, so e-mail was the first really popular invention to make use of computer networks. E-mail existed not because most people preferred to communicate in writing, nor because the technology was especially suited to writing (although for people like me, the letter-writing aspect of on-line communication helped smooth over some of the more alien aspects of the technology, and made it seem friendlier), but because sending written words required far fewer bits than sound or pictures, and in their primitive form most computers and networks could handle only small numbers of bits at a time. Voice was more complicated to express in bits than written words, and pictures required many more bits to express than sound, but in time technical progress would create networks and switches capable of handling these enormous quantities of bits, too, and perhaps when that day arrived people would look back at this brief platonic Age of E-mail with the nostalgia that we now apply to the brief Age of the Bicycle, before the noisy and awful-smelling internal combustion machines took over the newly built bicycle roads.

As the network grew, its members used it to thrash out the technical and social issues that its growth engendered. The syntax and nomenclature that seemed totemic by the time I arrived on the Net —the "@" sign between the user and the host in e-mail addresses; the "com" indicating a commercial domain—were casually hacked together by the pioneers. Users debated free speech, privacy, and the ethics of a new form of on-line discourse known as "flaming" —a kind of overheated style of arguing that was unacceptable face to face, or even on the telephone, but which computer-mediated communication made possible. To facilitate these and other discussions among network users, mailing lists were created. Usenet, a distributed bulletin board system of "newsgroups" to which readers posted "articles," began in 1980, and quickly grew to include every imaginable discussion topic. In the mid-1980s the National Science Foundation built a new, cross-country network backbone, with higher speed lines linked to faster supercomputers. ARPAnet and NSFnet linked to each other. Federal agencies started linking their computer networks to the NSF backbone. The backbone was beginning to branch off in different directions and become quite hairy with nodes. At first people began calling the Internet by its technical name, "an internet," meaning a network of networks. Eventually the word was apotheosized into the upper case, although the lower-case spelling was preferred on the Net itself; it was a kind of pledge of allegience to the Net's every-node-is-equal roots.

Through the 1980s the engineers and computer scientists who were the early settlers of the Net were joined by life scientists and researchers from the National Institutes of Health, geologists, physicists, and mathematicians, who were then joined by professors of literature and economics. Local area networks established by businesses for their employees began linking to *the* network. Independent networks that were set up for sports fans, musicians, pornography enthusiasts, writers, seniors, gays, neighbors, and white supremacists linked to the network. In the mid-1980s, the Big Three commercial on-line networks—CompuServe, America Online, and Prodigy—made e-mail and other information services available to the general public. Originally they were separate networks, but by the time I arrived all of them had e-mail access to any address on the Net, and were busy creating other forms of connectivity. The Big Three were the on-line equivalent of the transcontinental railroad of the late nineteenth century, the route by which the masses of ordinary settlers got out to frontier. There were also trade networks, like Sonicnet, which was an all-music information network; muncipal networks, like the Freenet in Cleveland, that provided low-cost community access; and neighborhood networks, like ECHO, which was based in Manhattan, and the WELL, in the Bay Area of California. All of this brought to the Net lawyers, journalists, college students, scam artists, lonely hearts, Holocaust revisionists, gun nuts, faddists who were buying CB radios in 1975, and me. It was like a whole generation of emigrants arriving in the West all on one day.

IV. The Many-to-Many ◊ ◊ ◊ ◊ ◊ ◊ ◊ ◊ ◊ ◊ ◊ ◊ ◊ ◊ ◊ ◊ ◊ ◊

The Internet was a "many-to-many" medium. The phrase was derived from the distributed structure of computer networks, in which every node was equal to every other node, but it was used as a metaphor for free speech and social equality in the new, networked world. John Gilmore, a former hippie who was now a wealthy software entrepreneur, had expressed the metaphorical relationship between network architecture and free speech in a popular aphorism: "The Net interprets censorship as damage and routes around it." Computer networks appeared to give the lie to a famous line of A. J. Liebling's: "Freedom of the press only applies to those who own one." Now anyone with a computer and modem was the owner of a printing press and a distribution network. You didn't have to formulate your message in language suitable for a Letters to the Editor page in order to get it published; you didn't have to go to the trouble and expense of buying time on a radio or TV channel in order to be heard; and you didn't have to pass out copies of your manifesto on a street corner. You could simply write your message on your computer, push a button, and publish it on a computer network.

People took advantage of Usenet to spread ideas about the Net's social potential.

hauben@cunixf.cc.columbia.edu (Michael Hauben)

Inherent in most mass media is central control of content. Many people are influenced by the decisions of a few. Television programming, for example, is controlled by a small group of people compared to the size of the audience. The ideas that exist on Usenet come from the mass of people who participate in it. Instead of being force-fed by an uncontrollable source of information, people set the tone and emphasis on Usenet. People control what happens on Usenet. In the tradition of amateur radio and citizen's band radio, Usenet News is the product of the users' ideas and will. Unlike amateur radio and CB, however, Usenet is owned and controlled solely by the participants.

eli@east.sun.com

The Internet represents the greatest decentralization in the dissemination of information since the printing press. We are about to undergo another information phase shift that will enable the people to talk to one another without the biased filters of the national media. Pretty soon, the talking heads that decide our national agenda and tell us how we feel about it will be lost in the clutter of *real* American voices.

Statements like these were heard in Philadelphia in 1776 and in Paris in 1798, except then they were made in favor of the printing press and the related technologies of typography and mechanized papermaking. Journalists like John Wilkes, Sam Adams, and Tom Paine were "tribunes of the people," much like the writers on Usenet were now. "Great is journalism," wrote Thomas Carlyle in *History of the French Revolution,* for "every able editor is now a ruler of the world." But on Usenet, print journalists were no longer the heroes; they were the villains now, the corrupt keepers of information that the many-to-many were about to liberate.

Print, according to this point of view, was a one-to-many medium. The information flowed from the publisher out to the masses, and it flowed one way. In many-to-many culture there was no hierarchical structure in the distribution of information, and the information could flow two ways. From the point of view of the many-to-many, a one-to-many system was elitist, because it created a structure in which a relatively few people were responsible for collecting and conveying the information, in a world in which information was becoming more valuable every day.

From the one-to-many's point of view, on the other hand, the many-to-many was the barbaric yawp from which the civilized technologies of print and broadcast had emerged. Inherent in the principle of one-to-many was the notion that some people were more worth listening to than others, and that the role of civilized society should be to make sure those exceptional people were heard, while the stupid people were kept as quiet as possible, so that everyone could take pleasure and wisdom from the voices of the few truly gifted individuals—the one(s). It was not a perfect system, but you had to set it beside the even greater imperfection of a many-to-many system, in which all the people with absolutely nothing to say, who had no training as newsgatherers and researchers, and no real accountability for getting the facts right, would have just as much claim on your time and attention as the interesting and accurate speakers. Wasn't many-to-many just a fancy phrase for the mob, the crowd?

But the weakness in the one-to-many argument was that it seemed to contradict what the new technology was saying, and in an age of technological determinism this appeared to be a grave flaw. As Carver Mead said, "Listen to the technology and find out what it is telling you." The technology seemed to be saying that we are all nodes on a network, flashing our electrical impulses to each other, with no one node greater than any other. It had become part of the framework of the argument that the technology was right, and to be on the side of the technology meant that the future would be your friend.

Paul Saffo, a futurist writing in 1993 in the newly launched *Wired* magazine—which was itself a hotbed of the new techno-idealism —speculated that many-to-many might cause a flowering of creativity and personal expression not seen in our society since the Sixties, and when that happened, sometime before the end of the decade, people whom we used to think of as computer nerds would have the same hipness that in nostalgic retrospect we now assign to beatniks. The central tenet of hippie romanticism—"Do your own thing"— was resurrected by Baby Boomers with computers. In a world in which people could download high-resolution images of paintings on flat panel display screens, why did you need museums? In a world in which people could download books, why did you need libraries? In a world in which you could download the music encoded on CD's, why did you need music stores? In a world in which authors could publish their works on-line, why did you need publishers? In a world in which you could get financial information from networks and execute trades on-line, why did you need stockbrokers? What was a community if it was rendered in bits and distributed through computer networks? What was "identity" in a many-to-many world?

> The technology of these transformative systems fulfills a profound human desire to transcend the limits of the human body, time and space; to escape language, to defeat the metaphors of self and identity that alienate and isolate, that imprison the mind in solipsistic systems. Our need is to fly, to reach out, to touch, connect—to expand our consiousness by a dissemination of our presence, to distribute self into a larger society of mind.

A vote was an exchange of information which could obviously be transacted on computer networks. But giving people the capacity to vote with their computers was only the first step. Why vote for politicians to vote for you on issues like abortion and balancing the budget, if you could vote yourself? Computer voting would make it easier to vote in general, which in turn would make democracy more participatory, not to mention more fun.

Tocqueville had commented on Americans' love for forming "associations":

> Americans of all ages, all stations of life, and all types of disposition are forever forming associations . . . religous, moral, serious, futile, very general and very limited, immensely large and very minute. Americans combine to give fetes, found seminaries, build churches, distribute books, and send missionaires to the antipodes. Hospitals, prisons, and schools take shape in that way. Finally, if they want to proclaim a truth or propagate some feeling by the encouragement of a great example, they form an association.

One could now add to that list "newsgroups," "bulletin boards," "chat rooms," "mailing lists," and other on-line gathering places of the many-to-many.

However, champions of the new medium feared that the one-to-many would not permit the many-to-many to remain free, but would try to limit or censor or regulate it. The many-to-many posed a threat not only to the authority of journalists and broadcasters, but to all forms of authority, because it permitted people to organize, think, and influence one another without any institutional supervision whatsoever—or so it appeared. How free should cyberspace be? Should there be a sovereign republic of the many-to-many, or should cyberspace be folded into existing constitutions, like beaten egg whites into a waffle batter? And how did you preserve the virtues of the many-to-many while at the same time imposing the hierarchy of government on it? Even in the United States, where constitutionally guaranteed civil liberties were much more compatible with the principles of the many-to-many than was the case in China, say, or Iran, the many-to-many would be difficult to work into the government. How did you create a Department of Cyberspace?

And all of this uncertainty and hope and fear was taking place in the shadow of the Cecil B. DeMille–sized calendar year 2000, which gave these changes the aspect of prophecy. Harder than the passage of a camel through the eye of a needle would be the transformation of Old World institutions into the new world of bits. A painful rendering was in store for many people.

V. Mysteries of Skill ◊ ◊ ◊ ◊ ◊ ◊ ◊ ◊ ◊ ◊ ◊ ◊ ◊ ◊ ◊ ◊ ◊ ◊ ◊

I answered all my e-mail about the Gates article as thoughtfully as possible. But I was getting fifty e-mails a day for the first couple of weeks, and worried about being overwhelmed. When, that first weekend, we went up to visit Lisa's mother in Albany, I took my computer with me, thinking, If I fall behind on my e-mail I'll have a hard time catching up. As soon as I walked into the hotel room where we were staying, my eyes started moving around the walls, about a foot up from the carpeting, not admiring the faux Colonial American wainscoting but hunting for an RJ-11 outlet (the standard telephone jack) that I could plug my modem into.

I tried to write e-mail in batches. I'd start the morning in front of my screen by composing replies to the eight or ten messages I had received during the night, then choose "Send and Receive Mail" from the Mail menu, which sent my replies, and pick up another eight or ten or fifteen messages, which I read over once quickly. After lunch I answered those messages, chose "Send and Receive Mail" again, and picked up another ten messages, which I answered and sent before going to bed, receiving at the same time another batch of e-mail for the morning. If I maintained my energy and focus I could answer forty e-mails in about two hours, spread over three writing sessions. It was like meditating three times a day. Composing e-mail composed me.

Previously disconnected ideas that had been sitting around in my brain for years began dropping into the flow of thought dribbling across my screen, creating interesting new patterns. Was this flow what George Gilder and other techno-utopians called "Mind"—a single entity emerging out of the chaos of the many-to-many? There seemed to be a new kind of energy flowing into my body through my fingertips, more like electricity than speech, which I would store inside my head for a while until it began to glow, and then send it out again through my fingertips and into the Net.

After one e-mail writing session, watching as my PowerBook distributed the messages in my Out Basket to points around the globe, I remembered Christmas Eve in the Deerfield Presbyterian Church, when we used to light each other's candles while singing "Silent Night": the light went from the minister's candle to the four ushers' candles, to the candles at the end of each row, to all the candles in the congregation. The spiritual aspect of the candle lighting was impossible to separate from the fact that this was the most efficient way to light a roomful of candles.

On another weekend that winter I went camping in the Adirondacks with some friends. There were no RJ-11 jacks in the woods where we were going, so I left my PowerBook behind. I thought that this encounter with nature might help to put the new technology into better perspective, but found that at night, snug in my nylon-and-down-and-polypropylene cocoon, I lay there listening to snow falling on the tent and thinking about how much e-mail I would have waiting for me when I got home. Getting up early the next morning, I walked into the woods, which were still and snow-covered and completely trackless. Casings of ice coated the tree branches. But the hum of the busy man-made world was buzzing inside my head. I had a vivid sense of men coming into diners somewhere outside of the woods, of the scrape of a metal snowplow on asphalt, of the squeegeeing sound of a child's hand rubbing a peephole in a misted-over classroom window.

Every morning that winter I sat down in front of my new toy with a sense of adventure. What new or old acquaintance would this day bring? Long-lost relatives came back from the past. Old friends, old schoolmates, old girlfriends, writing in from the four corners of the world. A woman who had lived down the street from my mother fifty years ago in Kansas City found my e-mail address and sent me mail, and I passed it along to Mom. "Good grief," my mother said, "I haven't seen that woman since 1947! It's amazing!" New friendships were born on e-mail, flourished briefly, then died and were replaced by others. Voices, voices—my computer was full of voices. Chance connections happened constantly, but at a level deeper than coincidence, it seemed to me—down at a level the nineteenth-century transcendentalists called "the eternal courrespondence of all things."

There were moments when it was possible to believe that I was feeling the same energy that flowed through the Black Hills of South Dakota in my great-grandfather's youth, down in the gulches where the placer mines were, where the people were burning up with it, and that now was flowing through the brains, nerves, plastic, silicon, and bits of the millions of people and machines on-line. This was the feeling that made you say, Yes, this is the place. But the place was hard to hang on to; it was always falling away from you, disappearing like the vanishing point of light on the old black-and-white TV sets. . . .

VI. **No E-mail from Bill** ◊

One reader (73631.744@compuserve.com) wrote to say that she had
sent e-mail to Bill, asking what he enjoyed about the novel *A Separate Peace,* and adding,

> Your job is enormously stressful. I really hope that you do yoga
> or tai chi or something else to calm your mind and keep you in
> the here and now.

She included <billg>'s reply:

From <billg@microsoft.com>
_____

No yoga yet.
   A Separate Peace is about growing up and finding your place
in the world. It deals with innocence and a youthful intelligence
trying to find its way.

I felt some envy, reading this message. I myself had received no
more e-mail from Bill since the article came out. Journalists are
supposed to operate on the assumption that if they don't hear back
from the subject of an article they must have done something right,
but I found myself wanting e-mail from Bill. In a *New Yorker*
column I did around this time about all the e-mail I got about the
article, in which I published some of the responses, I wrote, "I
almost crave it." By making it so easy to communicate with people,
e-mail changed the nature of communication; but e-mail, I was now
discovering, also changed the nature of silence. The silence of no
e-mail was unlike the silence of a quiet telephone or an empty
mailbox. It was thunderous.

Chapter Four **My First Flame**

The invisible worm
That flies in the night
Of the howling storm . . .

William Blake

I. **Crave This** ◊

In March 1994 the weather warmed up enough for the ice on the
ledges outside my room to melt. On the first Friday of the month, a
windy day, I carried my PowerBook into the *New Yorker* to do some
work, turned it on, immediately checked my e-mail, and saw I
had messages from two technology writers who collaborated on a
biography of Bill Gates. I had sent some e-mail to one of them
about getting together when I came out to Seattle to interview Gates,
and he had replied that it sounded like a good idea; however by the
time the interview had rolled around, I decided not to get together,
which was fine with him, too. I had used their biography as one of
my sources for the piece (there were two other, earlier biographies
that I had relied on more), and had acknowledged (I thought) my
debt to them by putting the name of the book and its authors into
the paragraph where their research appeared, in a way I thought
would help promote their book. In my trusting, doglike naïveté, I

actually thought they were writing to thank me for giving them a plug, and as I clicked on the "Open" icon I had already begun to compose in my head my friendly reply.

The first message said:

Crave this, asshole

Listen, you toadying dipshit scumbag, I'm going to assume your understanding this stuff is beyond you. But next time make at least a minimal effort to get your facts and spelling straight. For extra credit, try citing your sources and ripping them off accurately.

Finally, remove your head from your rectum long enough to look around and notice that real reporters don't fawn over their subjects, pretend that their subjects are making some sort of special contact with them, or, worse, curry favor by TELLING their subjects how great the ass-licking profile is going to turn out and then brag in print about doing it.

Forward this to Mom. Copy Tina and tell her the mag is fast turning to compost. One good worm deserves another.

The second message was more succinct:

I fully concur with [my colleague's] rude and angry words. What a rank shithead you are.

I rocked back in my chair and said out loud, "Whoa! I got flamed." I knew something momentous had just happened to me, and I was waiting to find out what it would actually feel like. I felt cold. The flame seemed to put a chill into the center of my chest which I could feel spreading slowly outward. My shoulders began to shake. I got up and walked quickly upstairs to the soda machines, then came back to my desk. There were the flames on my screen, not dying away like insults shouted in the street, flaming me all over again in the asynchronous eternity that is time in the on-line world. Being premeditated, the insults had more force than insults shouted in the heat of the moment, and the technology greased the words— the toads, scum, shit, rectums, assholes, compost, and worms—with a kind of intelligence that allowed them to slide more easily into my mind.

It seemed as though in that moment I understood the novelty and the power of the technology I was dealing with for the very first time. What was a flame but a perfect use of the medium—a sinuous ripple of emotion from brain to nerve to plastic to silicon to bits? Freedom of speech! Of course! The same intimacy that inspired people to speak their hearts also inspired them to say things that would be literally unspeakable in any other medium. Words that could not be spoken to my face (I would have walked away, or yelled back, or charged) or on the telephone (I would have hung up) could be typed and sent through the computer, and since the technology had made this form of human expression possible, people were naturally going to take advantage of it.

I printed the flames. On paper, they were less intense. They seemed to require the glow of the screen for their venom. I showed the printed copies to people around the office. Women were sympathetic; men advised me to grow a thicker skin. When I tried to explain my flamed feeling to one non-computer-using woman, she said, "Yeah, it's like when someone breaks into your car," which was close, but actually it was more like someone had broken into my head. It was the same feeling of being wired into each other's minds that I enjoyed in e-mail, but this was not enjoyable at all.

I returned to my office. I felt creepy sitting in front of my screen, like it was watching *me* now. I tried to get busy with other work in other windows, but my mind kept drifting over to the window containing the flames. Were these fellows respectable Jekylls of print journalism by day who changed into flaming Hydes of ASCII when they got home at night? Clearly something about the Gates article had pissed them off. Perhaps the flames were partly fueled by intense hatred of Gates, or Gates's products? After all, I was a Mac user. I had never had to endure Windows or DOS. How many awful frustrations, how many crashes, how many cumulative hours spent waiting on the Microsoft help line, were boiled down into these flames? But it was also true that these guys seemed to have a little problem with . . . *moi.* I guess a person like me—who obviously knew little about the technical components of computers, and who admitted as much in his pieces, and whose ignorance was made even more plain by a couple of embarrassing boners in the Gates piece, like confusing memory with hard disk space—could annoy the hell out of people who had been writing about computers for years. Worse still, my pieces seemed to suggest that I didn't think that the technology itself was so important. (True. How could you be a great technology writer and not keep in mind that the technology doesn't matter, ultimately—that a technology is immature until you no longer notice it?) What I seemed to be really interested in was not computers but . . . FEELINGS! And WORDS! WHAT A DIPSHIT!

Yes, I could see how someone might think that. But the fact that someone would actually say it was remarkable. You might conceivably write such a letter, but somewhere between addressing the envelope, licking the stamp, and walking to the mailbox, the heart for it would go out of you, or socialization would kick in, and you'd toss the letter into the trash. Yes, hate mail does exist in paper form, but it comes from anonymous readers, cranks, and wackos— not from people who are connected to you in some professional way. E-mail had legitimized hate mail for people who would not think of sending hate mail in the traditional sense.

As I was working that morning, every ten minutes or so I would stop what I was writing and start typing the most savage, hateful insults to my attackers that I could dream up. Anger, hurt pride, fear, and a lust for mayhem had all joined forces just below the calm surface of my superego, and were sending e-mail to the better angels of my nature which said, Put me in, coach, I'm ready to play. The insults went right into the draft of a piece I was doing on the Human Genome Project, like an attack of Tourette's Syndrome in the fingers. A toxic ripple of pure venom would flare up inside me, like a little flame fed with a jet of oxygen; I had energy and complete focus in composing these insults that I only wished I could muster for the piece that I was writing. Then the terrible immediacy of e-mail came over me and I had to fight hard to stop myself from sending what I had written. What *was* stopping me? Some notion of "decency," I suppose, but what was that? Wasn't "decency" really just a code word I was using to sneak some of my one-to-many values into the many-to-many world? The Net was not the Princeton Club, after all, and it was not the *New Yorker* either. (The roof of the Princeton Club was only a well-flung eraser away from my window at the magazine.) I leaned back in my chair, my finger resting on the <Return> key, thinking to myself that with a slight downward pressure of that finger my flame would be fired into their kneecaps, and I would have said something I had never said to any other human being before.

But wait: what if the flamers were right? Maybe I had been too soft on Gates. I didn't think that my portrait of his nondigital being *was* attractive. (Shortly after the piece appeared, *Time* magazine did an item noting similarities between my description of Gates's mannerisms and Oliver Sacks's description of the symptoms of autism.) But <billg> had indeed come across in an attractive way, as I had predicted, and for that perhaps I did warrant a good flaming.

I dragged my cursor across my insults, chose <Copy> from the <Edit> menu, hit the <Delete> key, and pasted the insults into a new window, which I had named "Insults File," and then tried to get back to work.

When I got home that evening I asked Lisa to look at some of my choicest insults. It was our first moment of togetherness in front of the small screen. She said, "Oh, honey, you can't say that. That's really terrible."

"It's no worse than what they said to me."

"I know, but still."

"Hey, it's free speech. If it's all about letting your emotions hang out, then I'm going to let mine hang out, too."

"Maybe you should just ignore them."

No. I had to say something. "A woman at the office suggested, 'Do you know where I could get a good bozo filter?' "

"That's good."

"Or how about something like this . . ." I typed:

Thanks for your advice on writing and reporting. The great thing about the Internet is that a person like me can get useful knowledge from experts, and for free.

We decided on that one. I dragged my cursor across it and pasted it into a blank message window inside the CompuServe software. I added another short paragraph asking for permission to pass along the flames to the Letters column editor at the *New Yorker.* Then I hit <Send>. In a few days I received replies from both flamers, denying the *New Yorker* permission to print their flames in the Letters column. One asked when my column "Pudlicker to the Celebrated" was going to start.

The evening after my flaming, two friends came over for dinner and I tried to describe what had happened. They asked to see the flames, so I went down the hall to print copies. But when I opened the electronic file in my computer where I store my e-mail I noticed that the title of my reply to the flames had turned into gibberish—instead of the letters there were little boxes and strange symbols. I also saw that the dates for when the message was created and modified said 8/4/72 and 1/9/04. This was strange, but I didn't think it was anything more than a software glitch, and when our guests had gone I put the computer into Sleep mode and then went to sleep myself. However, at just before six on the following morning I awoke with a start and sat up in bed with a sudden understanding of what the last line in the flame, "One good worm deserves another," might mean. A worm, as I understood it, was a kind of computer virus. One good worm deserves another: what if my flamers had also sent me a worm!

I got out of bed and went down the hall, woke up my computer, opened my e-mail file, and saw with a little tremor of fear that the corruption had spread to the title and dates of the message stored next to my reply. The reply itself was still corrupted but the gibberish and weird dates had mutated slightly. I tried to delete the two corrupted messages, but the computer told me it couldn't read them. Fear began to buzz in my chest. I copied the whole file onto a floppy disk, removed it from my computer, dragged the original file into the electronic trash can, emptied the trash, and then sat there regarding my screen with a mixture of suspicion and fear. I had the odd sensation that my computer was my brain, and my brain was ruined, and I was standing over it looking down at the wreckage.

I noticed I was sitting in the dark, so I got up and pulled the chain on the floor lamp, and the bulb blew out. I knew that the thought was completely illogical, the very opposite of the kind of thinking a computer would do, but it went through my mind like a virus nevertheless—wait a second, if my computer is connected to the electrical network in our house via the plug, is it possible that the worm could have jumped down my power cord, gone into the wall, and come out again in the lightbulb?

The worm had entered my mind.

II. **The Worm Returns** ◊ ◊ ◊ ◊ ◊ ◊ ◊ ◊ ◊ ◊ ◊ ◊ ◊ ◊ ◊ ◊ ◊ ◊ ◊

On Monday morning I took the floppy disk with my damaged files on it into the office to show it to Dan Henderson, who had set up the network at the *New Yorker.* Dan was a systems guy. Every office where the computers were networked together had a guy like Dan around, who was usually the only person who really understood the system, and these guys were often terrifically overworked, because in addition to doing their jobs they had to deal with all the people like me who were mystified by or actually frightened of their computers. Like many of the "word people" around the *New Yorker,* I was somewhat intimidated in talking to Dan about computers. I tended to agree enthusiastically with everything he said.

I sent Dan a QuickMail and told him that I thought my computer might have been infected with some sort of worm. I asked if he had time to see me, expecting that maybe he'd get to me before the end of the week. I was surprised to see Dan appear in my door within ten minutes.

"You QuickMailed me," Dan said. I noticed he was looking at me strangely.

"Yes . . ."

"You sent me QuickMail."

I was slow getting his drift. "So?"

Then I got it. "Wait. You mean you think I infected the office network?" Dan was just looking at me, his eyebrows up around his hairline. "But I took the infected files off my hard disk and put it on here," I said defensively, holding up the floppy.

He sat down at my computer with a couple of different kinds of software that searched for worms and viruses. I tried to follow his moves on my keyboard, hoping to pick up some shortcuts through the Macintosh operating system which I didn't know, but his hands moved too fast for me to follow them. After about ten minutes of probing he announced that he couldn't find any evidence of infection. He checked the floppy and found nothing there either. The gibberish and weird dates had gone away.

Dan looked at me patiently. I could see he thought I was somewhat insane. My worm fear was obviously not technical thinking. It was some kind of peculiar literary thinking that didn't make any sense. He explained that I could not have received a worm or virus via e-mail, because worms are programs, whereas e-mail is only text. A file containing a program could be sent over e-mail, but in order for it to infect your computer you'd have to open the file and run the program.

After talking to Dan I was persuaded that my worm was, in fact, just some weird software glitch that I had never seen before, and that it just happened to choose my reply to the flame to make its first appearance, and that the line "One good worm deserves another" was merely a strange coincidence. After thinking about this for a couple of days, I came up with a little experiment. My hypothesis was that perhaps the worm could have burrowed into the program I was using to set up a reply to the original message, and my experiment was to open the original message, click "Reply," and then file the document I had created with that operation.

The next morning my new reply and the message stored next to it were corrupted. I got Lisa's camera and took a picture of my computer screen. Then I called Dan at home.

"Dan? This is John. Dan, my worm is back. I'm looking at it now."

Dan was polite about it, but it was clear that he did not consider himself my network supervisor at ten o'clock on a Saturday morning. He said, "Could we talk about this on Monday?"

In the meantime, I had sent e-mail to the authorities at CompuServe, asking whether their subscribers were allowed to call each other rank shitheads and toadying dipshit scumbags. Eventually I got this reply:

To: John Seabrook
Fr: Dawn
Customer Service Representative
Since CompuServe Mail messages are private communications, CompuServe is unable to regulate their content. We are aware of an occasional problem with unwanted mail and are investigating ways to control such occurrences. If you receive unwanted mail again, please notify us of the details so that we can continue to track this problem.

I also sent e-mail to my computer-literate friend Craig Canine, a writer and farmer who lived in Iowa, asking what he knew about worms, and he replied:

From: <Craigk9@aol.com>

Coincidentally, I just gave our goats their worm medicine. It's called Valbazen, and it seems to work pretty well for ruminants —I'm not sure about computers, though. What does this worm do? Should I be communicating with you—might your e-mail be a carrier? Jesus, I've got my book on my hard disk. If your worm zaps it, I'll kill you first, then go after the evil perp (then plead insanity, with cause).

I sent a copy of my flame to a woman named Jennifer, a molecular biologist I had met on-line who had become a good source for me on the Human Genome Project article. She replied,

I must say that I was shocked to read about your experience. The magnitude of your assailant's tirade rends my heart. I've been thinking about those graphic words, unbidden, for the last two days.

You are right about the coldness of the Net. There is an air of preestablished hierarchy there—if you're new to the Net, or even to a particular group on the Net, you don't belong a priori. As a woman, I have encountered an additional barrier; the net is heavily male and women who want to play with the big boys either have to be ultra tough-talking—"one" of the boys—or else play off as coy, charming, "little-ol-ME?"-feminine. (Even geeks have fantasy lives, I suppose.) Or use a male/neutral alias with no one the wiser.

So part of the boys' club, I imagine, is the smallness, the selectivity—the geek elite, if you will. For more than a decade these guys had their own secret tin-can-on-a-string way to com-municate and socialize, as obscure as ham radio but no pesky FCC requirements and much, much cooler. But then the Internet —their cool secret—started to get press . . . Imagine these geeks, suddenly afraid that their magic treehouse was about to be boarded by American pop culture. It was worse than having your favorite obscure, underground album suddenly appear on the Billboard charts.

What a great e-mail! The flames had taken something out of me, but now I was getting something back.

Another on-line acquaintance sent me this story:

When I first connected to the Internet there was a period of intense lurking, an activity which isn't as passive as one might imagine; I would read from various discussion groups—usually the more chaotic and uncivilized ones—and then log out and practice my own incendiary diary responses to the comments I considered most inane. There is something brutal and predatory about much of what goes on on the Internet. There is a kind of smart-ass style one must either learn to ignore, or capture and exploit for one's own purposes. The law of the jungle seems the guiding impulse behind much of what is encountered, and being a novice I succumbed to its false clarion call. In one group, an individual appeared constantly as an informed and authoritative voice. He didn't engage in vitriol, yet he didn't shy away from vigorous debate and the cultured slight. His address suggested some sort of educational affiliation, possibly a university professor, though I couldn't tell absolutely. So I sent him an e-mail message in which I attacked his "personal style"; my attack was outlandish, crude and irrelevant to the content of any of his postings. As I typed, I felt a blinding excitement, for it seemed to me I had focused on a fiction, as one focuses on a toy duck at the carnival, and I assumed or thought that my volley would be received as a further fiction and a kind of duel would ensue, entertaining all innocently and intensely.

The next day I checked my e-mail and found only a terse acknowledgment that my message had been received and duly noted. It was then I realized the complete error I'd committed. The real hurt I might have inflicted, the deep distress. For it was the shortness of his response that caused me to shiver: what could I do to make it up? How could I right the wrong I had done?

I did write another e-mail in which I tried lamely to retract my boorishness, but this did not work as I quickly received a letter of outrage and disgust at my action. This pleased me somewhat; I reasoned a man willing to express outrage has certainly not been harmed or discouraged permanently as a result of my stupidity and ignorance. Yet I wasn't satisfied with what I had done, and I felt my apology was insufficient to the act. So I posted a public apology on the list where I first read his postings. This apology included no details other than the facts of my having sent an inappropriate message and a promise no such messages would ever be sent again, and that any further communications between us would be public or not at all.

Yet there is a final, rather sad irony to this small affair: my public apology was flamed by several participants in the list, and the person to whom I had addressed my apology made a guarded defense on my behalf.

I also forwarded a copy of the flame to my mother, as one of the flamers had suggested. She replied:

I deleted that thing you sent me immediately. What a terrible man. He must have been drunk.

III. **Welcome to the Internet!** ◊ ◊ ◊ ◊ ◊ ◊ ◊ ◊ ◊ ◊ ◊ ◊ ◊ ◊ ◊ ◊

I tried to think of someone I knew in real life besides Dan who might know something about flaming and viruses and all this other computercentric stuff, but couldn't. My on-line life and my real life were still almost entirely separate from each other. By now I had a few friends IRL (which was how "in real life" was written, on-line) who I also exchanged e-mail with, but we rarely talked about what we said to each other in e-mail in our f2f (face-to-face) conversations, and if we did, we spoke a bit sheepishly about our e-mail, as if we were referring to our secret affair.

It occurred to me that I could try going out onto Usenet and looking for a newsgroup devoted to these subjects. It is perhaps ironic that this most unpleasant experience actually drove me deeper into the on-line world, in an effort to understand it, but that is what happened. And at the same time, the practice of flaming, at least as it was employed by the Net dot pioneers, appealed to me. It was a rigorous, sometimes cruel, but usually earnest effort to search out truth and blast away cant and obfuscation without worrying too much about how one ought to behave. How would it be to live that way? To let your thought-dreams be seen! To say to your fellow citizens, Let's not allow social conventions to determine the way we speak to each other; let's just say what we think. That might be painful, but that might also be the path to enlightenment. At any rate, as I ventured deeper into the dark woods of the Net, it seemed as if flaming was the direction that the light was coming from, so naturally that was the direction I wanted to go.

In those days you couldn't get to Usenet from CompuServe, so I had to sign up with another service provider, Delphi, for access to Usenet newsgroups. After an hour of trial and error with Delphi's keyboard-command-style interface I got to the first layer of Internet menus. I had reached the frontier.

Suddenly my screen froze. What had happened? I knew whatever I had done wrong was a simple, logical mistake, but I couldn't figure out what it was. Feelings of helplessness and frustration spread into my mind like saltwater into a paramecium, exploding my capacity for reasoning. There is nothing in the world quite so deserving of one's contempt as a machine too stupid to do what it's supposed to do. I disconnected from Delphi by unplugging the wire from the back of the modem, then tried to connect again, and got a message saying <Unable to Initialize Modem>. I quit the communications program, opened it again, and got the same message. I restarted my computer, logged on to Delphi again. This time: Success.

The pioneers had divided Usenet newsgroups into hierarchies— "alt," "rec," "comp," "sci," and "soc"—which stood for "alternative," "recreational," "computer," "science," and "social." The

Delphi newsreader gave me an alphabetical list of them, beginning with the "alts." I began scrolling, and soon came to one called alt.flame. But alt.flame turned out to be a place where people went to flame each other.

Hi! You probably wonder why I've not written to alt.flame in awhile, it's because I am in Finland, I hope you're all miserable. FUCK YOU!!

Hey, we were having a party. That was till you showed yer fat ugly face again.

A lot of the flaming in alt.flame was just a game, like a white, nerdly version of "The Dozens." For instance, I saw a posting entitled FLAME WARRIORS! Read IMMEDIATELY!!!!

Greetings flame warriors.

Do you cruise the Internet from time to time in search of your next victim? Have you ever said to yourself "hmmm, let's look at the list of newsgroups and find a group of people who deserve my wrath." Do you enjoy blasting to bits the wills and the self-images of your fellow Net-surfers (you know, the fools who stupidly incur your wrath). Are you armed to the teeth with pointed barbs, quick-witted replies, devastatingly destructive posts, and the like?

Most importantly, do you have the skills of wit, creativity, originality, intelligence (or more precisely—brilliance), confidence, and the will to use them for your own personal pleasure?

If you do, then you are a Flame Warrior. Certainly, you may have felt that it would be nice if powerful people like yourself could congregate and communicate with each other. Wouldn't it be grand to forge a "Round Table of Noble Flame Warriors." Certainly there is little as noble and as deservedly praiseworthy as the ability to flame with unimaginable intensity!

Well, it is this person's ardent desire to unite the currently fractious cadre of my Flame Warrior brothers and sisters. As a united force, we would command even greater respect and honor. We could bear our status as Flame Warriors with pride. Our presence would evoke not only fear, but awe and respect.

But sometimes the game could get a little nasty. I saw that a writer from *Wired* magazine, Amy Bruckman, had recently posted that she was researching an article about flaming, and had asked for people's views on the practice, and was now getting some "information" back.

Insert finger in appropriate orafice and shove off.
Sod off bitch, we don't need your glamour here. . . .

WHAT?!? Do you think I wanted to be publicated in your low-life-scum magazine??? . . . BTW, what kind of name is Bruckman? Are you kind of a German refugees' daughter from the 2'nd world war. Kraut! a sauceage woman? Anyway go to hell.

Now I just leaned back and went surfing, an experience I was to repeat many times in the months ahead. I moved on through the newsgroups, got distracted by a group named alt.pagan, stopped there and read the FAQ. The FAQs, or Frequently Asked Questions files, were the repositories of useful knowledge and meaningful events that had occurred in that newsgroup since it was established —the culture, basically—and were usually kept by volunteers. The table of contents of the alt.pagan FAQ was organized with a clarity and rigor that would have pleased Aristotle.

1) What is this group for?
2) What is paganism/a pagan?
2b) What is Paganism? How is it different from paganism?

3) What are different types of paganism?

4) What is Witchcraft/Wicca?

4b) Why do some of you use the word Witch? Wiccan?

5) What are some different traditions in the Craft?

6) Are pagans Witches?

7) Are you Satanists?

8) What kinds of people are pagans?

9) What holidays do you celebrate?

9b) How do I pronounce . . . ? What does this name mean?

10) What god(s) do you believe in?

11) Can one be both Christian and pagan?

12) What were the Burning Times?

13) How many pagans/Witches are there today?

14) Why isn't it soc.religion.paganism instead of alt.pagan?

15) Is brutal honesty or polite conversation the preferred tone of conversation around here?

16) What are the related newsgroups?

17) Are there any electronic mailing lists on this subject?

18) I'm not a pagan; should I post here?

19) How does one/do I become a pagan?

20) What books/magazines should I read?

21) How do I find pagans/Witches/covens/teachers in my area?

22) What's a coven really like?

23) How do I form a coven?

24) What does Dianic mean?

25) Aren't women-only circles discriminatory?

26) Can/will you cast me a love spell/curse my enemies?

27) Is it okay if I . . . ? Will I still be a pagan if I . . . ?

28) I am a pagan and I think I am being discriminated against because of my religion. What should I do?

29) What one thing would most pagans probably want the world to know about them?

Had I been a "Broom Closet Pagan" (a phrase I picked up in alt.pagan), the thoughtful answers to these questions might have brought me out of the closet. Were you supposed to come out *on* a broom?

I dropped smoothly down to the next level of topics, feeling a flush of pleasure at having added this little bit of software knowledge to my medicine pouch, and read the alt.pagan discussion topics.

14 European paganism (16 msgs)

15 Statement (6 msgs)

16 College Pagan Groups

17 PAGAN FEDERATION GIG: Thanks (3 msgs)

18 Broom Closet Pagans Hurt Us All (3 msgs)

19 Pagan funerals? (27 msgs)

20 NIGGER JOKES (18 msgs)

21 Necromancy (2 msgs)

22 Another campus Pagan group (4 msgs)

23 When the Revolution comes was Re: New Forest Service . . .
      (6 msgs)

24 looking for invocations to the following . . . (4 msgs)

25 New Community Pagan Group? Need help

"NIGGER JOKES"? What did they have to do with paganism? The posting was a collection of jokes and limericks about killing black people. It had been "spammed," as they said on the Net, which meant it had been spread around willy-nilly to many different newsgroups, to an audience of tens of thousands, maybe more, depending on how many people copied the jokes and passed them along. The name and address on the jokes was a student at the University of Michigan, and he was now getting flamed hairless in alt.pagan. I could only imagine what his e-mail must be like. A few people had called for him to lose his account, and these people were now being reprimanded by others in the topic, for trying to impose censorship on the Net—the recipe, I was beginning to learn, for the classic Usenet flame war.

The Last Viking <paalde@stud.cs.uit.no>

We don't have to go around being racists like those fascists in the real world! PEACE ON THE NET!!!!

Michael Halleen <halleen@MCS.com>

He should get hate mail, censure (not censors), and universal condemnation. There should be open debate and discussion, but leave his right to speak alone. He may use the net for other constructive purposes and taking it away may hurt him, and he needs help.

Richard Darsie <darsie@eecs.ucdavis.edu>

Get a grip, man. Free speech is not and never has been an absolute right. There's gotta be some limits. . . . This person abused his First Amendment right and should face some consequences for it. Can't have rights without responsibilities.

Jessica M. Mcgeary <jmm25@po.CWRU.edu>

Anyone care to ask root or postmaster or whatever why this fellow has been allowed to post this?

Jeffrey B. Deutsch <jdeutsch@mason1.gmu.edu>

Repeat after me: "Freedom of expression—even for those expressions that we hate, freedom of expression—even for those expressions that we hate, freedom of expression . . ."

Reading through the postings, I came to the information that the student whose name was on the address had not, in fact, been responsible for posting them. The wrongly accused student, now perhaps flamed beyond recognition, had used a university-owned computer to log onto his account, and someone had tampered with the software in that computer so that it captured his password. The tamperer was the person who had actually posted NIGGER JOKES, and was no doubt thoroughly enjoying the flame war his post was causing. And perhaps was participating in it, under another identity.

On Usenet I found "associations . . . religious, moral, serious, futile, very general and very limited, immensely large and very minute," just as Tocqueville found on his tour of the United States. There were friendly places like rec.birds, for example, or rec.crafts.textiles.quilting, or alt.lifestyle.barefoot, where I rarely saw anything but heartfelt enthusiasm for birds, quilts, and the joys of walking around with no shoes on, respectively. And even where things were beyond the boundaries of polite society, this was the real American dial tone—a free people, speaking freely to each other, comin' at you on 1999.99 on your millennial dial. Freedom to write in the lower case. The freedom to channel some funky voice that was dancing around the edges of your consciousness into your postings. The freedom to invent new words. The Net was a hotbed of language, because language had to accomplish everything; the whole world was made of words. Because typing took more effort than speaking, there were lots of acronyms and abbreviations— "lol" for laughing out loud, "IMHO" for In My Humble Opinion, and "RTFM" for Read the Fucking Manual, which was a message people sometimes sent back if you asked them for technical help on-line.

I saw a lot of sex. In the newsgroup alt.binaries.pictures.erotica, for example, was this list of downloadable images:

2913 mbon001.jpg (1/1) {female} "Bondage—Two in armbinders, gags, nipples chained between ladies"

2914 mbon002.jpg (1/2) {female} "Bondage—Two cuffed together, gags, nipple clips"

2915 mbon002.jpg (2/2) {female} "Bondage—Two cuffed together, gags, nipple clips"

2916 mbon005.jpg (1/2) {female} "Bondage—Cuffed to A-frame, breasts bound, gag, collar, crotch rope"

2917 mbon005.jpg (2/2) {female} "Bondage—Cuffed to A-frame, breasts bound, gag, collar, crotch rope"

2918 mbon009.jpg (1/2) {female} "Bondage—Hogtied with 3 handcuffs, breasts bound, crotch rope"

2919 mbon009.jpg (2/2) {female} "Bondage—Hogtied with 3 handcuffs, breasts bound, crotch rope"

The Net was especially good for the nerdy pornography enthusiast-collector type of user who enjoyed rare pictures of bondage, coprophilia, bestiality, or some other particular flavor of porn, but who did not wish to go to the trouble and possible embarrassment of procuring these items in a store or through the mail. In the future, with progress and the bandwidth promised by the techno-utopians like George Gilder, it was possible to imagine having a private hard-core peepshow on your desktop, running twenty-four hours a day, only a mouse click away. And if that happened, it was easy to imagine some of the champions of this medium becoming its censors. For while progress may be a feel-good religion, it is not an especially open-minded, humane, or tolerant one. It can't really afford to be. There are too many obvious examples of lack of progress in the world, and at a certain point the evidence they supply is so overwhelming that you begin to wonder whether progress exists at all. But when you get an American with fantasies about the perfectibility of the world all excited about the existence of real progress through technology, and then it turns out that people are using this shining example of progress to do filthy things—well, there's hell to pay.

In many newsgroups there was rigorous argument, even the tearing apart of one person's ideas by another, but always as a way of getting at the truth; it stopped short of flaming. (A Nethead once described this form of discourse to me as "nerd jousting.") And then there were flame wars. The FAQ of some newsgroups were *Star Trek*esque tales of these flame wars, which were often provoked by a single poster going off on some jag and writing loonier and loonier posts until he/she seemed just to disappear into a crazy grin, like the Cheshire Cat in *Alice's Adventures in Wonderland.* There was L. Detweiler of alt.conspiracy, who devoted amazing energy to spinning out his conspiracy scenarios, and if you posted saying he was paranoid, he accused you of being part of one of them; Hannu Porupudas of sci.physics, who interpreted messages from God in the form of drawings made by his six-year-old daughter; Dan Gannon of alt.revisionist, who argued that no Jew, Gypsy, or other person was purposely killed as part of a national policy by the Nazis during World War II, and that there certainly were no mass killing operations, no gas chambers, no gas vans, no mass executions; John Winston, who wrote about alien visitors and UFO's, and who sometimes began posts with "Dear people who breathe air"; Dave Hayes, who so often claimed to be speaking for the "little people" of Usenet that someone eventually coined the saying "Speak up for yourself or Dave Hayes will speak for you"; "Lt Wilkes," who maintained that his brain and body were in two separate places, and that one of them had recently visited Pluto; and Daniel J. Karnes, who posted incendiary comments in alt.homosexuality, because, by his own account, "Early in his net.career djk noticed a strong homosexual element present on the net, and being a member of a very large Traditional Values group, decided to watch them closely. It did not take much time for djk to become outraged by the activities of the net.queers who seemed to think that the Net was their private playground. . . ."

One could try to respond to a troublemaker with humor and ridicule. One woman on a *Star Trek* mailing list wrote to a puerile antagonist:

> Your constant application of the epithet "rhinobutt" to a woman of whose radiant form you have never been vouchsafed a glimpse tells us less about your opponent than about your own gluteal obsession.

But a determined sociopath intent on mayhem would often succeed in plunging a newsgroup or mailing list into a flame war, and in some cases destroying it. It only took one jerk bent on exercising his free-speech rights to the max to turn any on-line group into a "spench pit" (another useful on-lineism, meaning a place polluted by speech), while at the same time acting within the principles of that free-speaking utopia. The good people in the group, unable to defeat the obnoxious poster on moral grounds, often had to turn to the backroom, authoritarian, one-to-many tactics that the many-to-many society deplored. As one poster on the com-priv mailing list put it:

> Now, I served as the national ACLU's telecommunications consultant for several years and will do anything I possibly can to preserve free speech, but I also recognize that the greatest threat to free speech is its wanton and deliberate abuse.

There were bad flame wars in the religious newsgroups, and the Turks and Armenians fought a long flame war in soc.history. Gun nuts and Aryan brothers laid waste to alt.militia. No matter what the origin of the conflict, many flame wars conformed to what was known on the Net as Godwin's law, which, as defined by Mike Godwin, on-line counsel for the Electronic Frontier Foundation, stated, "As an on-line discussion grows longer, the probability of a comparison involving Nazis or Hitler approaches one." Posters from alt.misanthropy and alt.tasteless came into rec.pets.cats and posted graphic accounts of butchering cats, and although the cat lovers could in theory "killfile" the miscreants after their first postings, so that they never had to read a message from them again, the shocking words tended to linger in your head—sooty black specks in the pretty picture of the many-to-many.

I had heard that in the newsgroup alt.fan.courtney-love, the rock star herself had been posting, so I went there, and sure enough, there she was, making a virtual version of one of her famous dives into the mosh pit:

I'm sorry if I've upset your concept of celebrity by actually participating in this, but HEY! your concept of celebrity was just boomer handed to you anyways . . . see, just like Thurston . . . I'M ACCESSIBLE! I LIVE IN A HOUSE! I HAVE A CHILD! I AM HUMAN! THE LIVING HIEROPHANT! THE COVER Of MS. AND ESQUIRE THE SAME MONTH! AND IT ISN'T A CONTRADICTION! I MUST BE. . . . WAIT FOR IT. . . . A FEMMENIST (Thanx Noam) okay fuckoff peace love empathy and kiss my ass . . .

Someone had replied:

The NIN crew says you were bombed every night you opened for Trent . . . and you embarrassed them, Trent, and yourself. The word is live, you suck. Good thing man invented the studio, ehh? G'head, honey . . . stop reading this and sink that spike. You'll forget all about it. Feels good, doesn't it? Remember when you didn't have any money and you were junksick all the time? Good thing you're a rich rock star, Courtney baby. Just OD already so Frances Bean can have a shot at growing up normal.

Some of the heavy metal music newsgroups were in complete chaos; people had been flaming each other for months, absolutely scorching everything around them, and driving all the civilized music lovers away. Sometimes I arrived at a dead site after a flame war had broken out; it was like walking through what had been a forest after a wildfire. I came across voices that were just howling at the world, rage and savagery pouring out through people's fingers and into the Net.

The ideal, of course, was to find a way to get the most out of the freedom that on-line discourse offered—freedom from shyness, isolation, gender, money, race, or whatever else you felt might be inhibiting your freedom IRL—without allowing the kind of abuse of free speech that laid waste to so many on-line groups: racists disrupting on-line groups of Afro-Americans, homophobes disrupting groups of gays, men trashing women. But the ideal was hard to achieve on Usenet. It was a rough place. You saw fear of getting flamed in many people's postings—long posts that ended, "Sorry so lengthy, please don't flame," and messages studded with smiley faces—:)—and grin signs—<g>—a bit like dogs cringing around each other to avoid starting a fight.

That spring I spoke on the telephone (a growing novelty in these days in front of the screen; e-mail had taught me I never really liked talking on the telephone; my throat tightened, the glottal equivalent of shy bladder) to Eugene Spafford, who was one of the earliest settlers of Usenet. His more recent efforts to get people to act with "decency" on the Usenet, and to exclude certain newsgroups about sex and drugs, had earned him the somewhat derisive nickname

"Emily PostNews." He told me: "What I see is new users getting on-line, promptly getting flamed by the older users, and then saying to themselves, hey, let's get down and dirty and go for it. In any other kind of medium, the reality of the two people talking would prevent a lot of ugliness from happening, but with nothing but bits between them, people feel they can say anything they like—it's not a human you're talking to, it's just a machine."

He said he had recently logged off Usenet for good. "I made my last set of postings about a year ago. I finally gave it up because not many people shared my view that there really needs to be a role for decency on-line. I just felt that the majority of people didn't want to go in that direction, or at least the majority of people who made their opinions known did not. A lot of old-timers like me are leaving the Usenet. And it's terribly frustrating because we know how great the Net could be. But I've decided to see what help I can be to my daughter and my students, and maybe in the long run I can do more that way."

After I had published an "article" in print about the flame wars I saw on Usenet, I received the following e-mail.

From: <acharles@umaxc.weeg.uiowa.edu>

Let me tell you something about flamewars. The undergraduate shitsmearing aside, they're serious things and ought to be allowed to continue. They are, consistently, about the most important problems we have. They cost almost nothing in real terms. They are the best streetlevel forum we have, and if they get violent, well, Jesus, they're about important things; of course they're violent. Verbally violent, and in a space where no fistfights, no (or rare) real-life vendettas can break out. What the hell else could you want, for hashing out these things? Examine the serious ones sometime. Look for where they break down, where they get circular, where people get tired and substitute half-remembered propaganda for reality. Look for the people who carry the arguments with consistently strong theses, with real vigor in their thought and understanding.

Forget the insults, the tactlessness, who the hell cares about it, it's not what's important, here. The dialogue. The engagement, the debate, the *willingness* to engage, and the real argument that drives the flame threads, *that's* what's important.

That was true. More than true, it was essential. But perhaps because I came from a family in which people fought with ice, not fire, the flames I saw in my early travels on Usenet shocked me, and discouraged me from posting in any of the newsgroups. Of course you could also hear flaming on the streets of New York City, but in the city the flames generally lasted only as long as the person who was making them had breath for them and were only heard by the people within earshot. On the Net flames could be heard by millions and reverberate for a long time.

IV. The Dark Side ◊

One day I asked Dan Henderson if he knew of someone I could go to for the final word on my "worm"—the top worm man in the country—and he gave me the e-mail address of a man named John Norstad, who was the author of Disinfectant, a popular brand of virus protection software for the Macintosh, and who Dan said knew as much about the viruses and worms that affect Macs as anyone in the world. I wrote e-mail to him explaining my problem, and he wrote back saying he was in the midst of fighting a new virus that had just broken out in Italy, and didn't have time to think about my problem now, but would be happy to talk to me when I came to

Chicago. Chicago was the site of the Computers, Freedom and Privacy Conference, an annual gathering of interesting netheads, privacy advocates, and defenders of civil liberties in cyberspace, which I planned to attend that year.

The CFP Conference of 1994 was consumed with the battle against the Clipper Chip, which was the federal government's first major effort to regulate the new medium of the many-to-many. The Clipper Chip was a fingernail-sized sliver of silicon designed by the National Security Agency, the ostensible purpose of which was to encrypt messages in such a way that only the person you sent them to would be able to read them, and which the administration was trying to sell to the public as a way of increasing personal security and privacy on the Net. The problem was that packet-switching made computer networks relatively insecure. Because packets of data might be routed through many different computers on their way to their final destination, it was relatively easy for "sniffer" programs installed in any one computer along the route to intercept your bits. This was not a problem if there wasn't much in your bits that anyone else wanted to read. But if people were really going to live on-line, if bits containing their medical records, their credit-card numbers, their bank balances, tax payments, and votes were going to flow through the wires, then the general insecurity of the bit-stream was going to be a problem. With Clipper, the government was proposing to solve that problem by making powerful encryption available to everyone.

But encryption also had a dark side. By making it easier to keep secrets, encryption would also make it possible to commit more perfect crimes. With powerful encryption, the Net would be an ideal place for criminals to organize conspiracies. If John Gotti had planned his crimes on-line, in a virtual Ravenite Social Club, and had he been using powerful encryption, there would have been no bugs and no wiretaps, and he wouldn't have been convicted. Dr.

Clinton C. Brooks, the NSA's lead scientist on the Clipper Chip project, told me that spring, "You wouldn't have a Waco in Texas, you'll have a Waco in cyberspace. You could have a cult, speaking to each other through encryption, that suddenly erupts in society— well-programmed, well-organized—and then suddenly disappears again." In an effort to make sure this didn't happen, Dr. Brooks and his team had built an electronic backdoor into the Clipper Chip that law-enforcement agencies could use, with a court order, to unencrypt the messages of people they suspected of being criminals.

The Net community agreed that powerful encryption should be available to everyone. But it advocated the use of other, nonofficial forms of encryption, like "Pretty Good Privacy" (PGP), the program developed by Phil Zimmerman and freely available on the Net. The Clipper Chip fight gave a moral urgency to the proceedings of the CFP Conference that year. The Clinton administration sent a hapless young man to the conference to try to defend its endorsement of the Clipper Chip, and the civil libertarians in the audience made mincemeat of him. The NSA after all was an agency whose main activity for the past forty years had been electronic surveillance—an organization so secretive that for many years the government tried to deny its existence. Putting spies in charge of protecting people's privacy on-line was bound to make people nervous. As everyone knows, the CIA has monitored groups that are not criminal, but that simply express unpopular political ideas. Given the choice between sharing c-space with criminals and being watched by government spies, a lot of people preferred to take their chances with the criminals. The government was reduced to using specters of on-line cults and conspiracies, and pedophilia, in order to scare people enough to support the Clipper Chip.

John Norstad and I had arranged to meet at my room in the hotel where the CFP Conference was held. As I was waiting for him to

arrive, I tried to imagine what he would look like. You could sometimes guess what people look like from listening to them talk on the
phone, but they almost never look the way they write e-mail. Norstad turned out to be about forty-five, not tall, with a beard that had
some gray in it, glasses, and a shy, polite manner. Although I had
formed a mental image of him dressed in the ceremonial robes of a
Sioux witch doctor, he wore a flannel jacket over a loose gray shirt,
and gray pants. He was carrying a PowerBook loaded with the
dominant strains of all the nastiest viruses known to the Macintosh
world; the viruses were safely corralled on his hard disk with his
program Disinfectant, which he distributed for free on the Net to
anyone who wanted it. Norstad placed his PowerBook next to my
PowerBook and showed me his collection of infected programs. He
moved his cursor over and pointed it at an icon, double-clicked on
it, and said, "Now if I didn't have any protection this little guy
would start erasing my hard drive right . . . now. But because we do
. . . there . . . see . . . Disinfectant caught it."

I asked about the Italian virus he had been fighting when I first
e-mailed him, and Norstad said it had appeared in a piece of software that was posted on a pirate bulletin board. Because the software was copyrighted, and had been posted on the board illegally,
there was some suspicion that the virus writer might have been
trying to teach the pirates a lesson about copyright infringement.
Norstad opened his e-mail log and showed me the several hundred
messages he had sent and received between February 28, when he
had heard via e-mail from three people in Italy who reported that a
new virus was erasing people's hard disks, and March 3, when he
and his colleagues had produced a vaccine. On hearing about the
Italian outbreak Norstad had immediately sent e-mail to a group of
colleagues called the Zoo Keepers, to alert them to the possibility
of the new virus. The Zoo Keepers were a virtual community that
lived all over the globe (Australia, Germany, the United States),

the kind of community that could only exist because of computer networks. Norstad received a copy of the virus from Italy, made copies, and sent the copies out over the networks to the Zoo Keepers. Keeping in touch by e-mail, the scientists reverse-engineered the virus and developed a number of effective vaccines for it. Norstad then updated Disinfectant—version 3.3 became 3.4—and posted it around the Net, where people could download it for free. All of this took fifty-six hours.

I asked whether virus writers were often motivated by politics, and Norstad said no, that they were mostly relatively harmless hackers, at least in the Mac world. In the world of IBM-compatible machines, there were many more viruses and they tended to be deadlier. Norstad told me of an account he had once heard from a Bulgarian virus expert, about software engineers commissioned by the Communist government to crack the security seals on Western software. When the regime fell, the story went, the unemployed engineers were said to have whiled away the empty hours writing viruses for IBM compatibles.

I told Norstad about my computer problems, and then asked whether it was possible to send a worm through e-mail. He confirmed that it was impossible. He added, "The kind of symptoms you describe could be a software problem."

"Like what?" I asked.

"Who knows?" Norstad said. "It's software. It's weird stuff. People are always writing and calling me because they think they have some kind of virus, and in almost every case it's a software problem, not a virus, but these people are fearful and need my help. For example, quite a few people have written me to say a shrieking death's head appears occasionally in the top of their screens. You know what it is? If you run Apple's Easy Access program, hold down the <Option> key, and hit the <Shift> key three times, your computer makes this funny trilling sound and an object appears in the corner of your screen that could, if you were sufficiently para-noid, look like a death's head. It's not a virus. It's just a weird software thing."

While Norstad was talking I had brought the flames up onto my screen, and now I asked him to look at them. He leaned toward my screen and silently read through the litany of insults. When he had finished he leaned back and sighed and didn't speak for a couple of seconds. Then he said, "I'm very sorry when something like this happens."

I said, "I have to admit that it was quite upsetting, and I don't really know why. I ask myself, Do I recognize the right of another person to flame me? Yes, I do. Okay. Do I recognize the right of that person to send me a worm? Definitely not. But at what point does a flame become a worm? I mean, can free speech sometimes become kind of like a virus? In other words, could a combination of words be so virulent and nasty that they could do a kind of property damage to your head?"

I could feel tears coming into my eyes, so I stopped there. I thought, Jesus, what am I getting so worked up about?

We chatted for a while longer, and then he said, "Look, don't get discouraged. The Net is a fundamentally wonderful thing. Most of this work I do could only be done on the Net. Look at the work we did on the Italian virus, working with colleagues all over the world to reverse-engineer it. Can you imagine trying to do this by fax? Phone? FedEx? It would not be possible."

Norstad chose Shut Down from the Special menu, unplugged his machine, and put it back in its case. "Of course," he added, "com-puter networks also allow people to spread viruses much more easily than was possible before."

"But that's the thing about the Net," I said. "Each of the good things about it seems to have an evil twin."

"Yes, but you could say that about all technology," Norstad said.

"There's always going to be a dark side to it. That's why it's so important to be decent on the Net, because the dark side is always right there."

As Norstad was putting on his coat, he said, "My thirteen-year-old daughter is a Pearl Jam fan, and the other night she asked me if there might be some Pearl Jam stuff on the net. So we logged on and looked around, and we were able to download some Pearl Jam posters, some music, some song lyrics—really neat stuff. But then we came to the Pearl Jam newsgroup, and there was a really terrible flame war going on in there. People were saying really awful things to each other, things I was embarrassed to be sitting next to my daughter reading." Norstad shook his head. "Terrible things."

Part Two **East**

It follows that the wise revolutionary legislator, so far from seeking to emancipate human beings from the framework without which they feel lost and desperate, will seek rather to erect a framework of his own, corresponding to the new needs of the new age brought about by natural or technological change. The value of the framework will depend upon the unquestioning faith with which its main features are accepted; otherwise it no longer possesses sufficient strength to support and contain the wayward, potentially anarchical and self-destructive creatures who seek salvation in it.

Isaiah Berlin, *Four Essays on Liberty*

Chapter Five   I Am a Node on the Net

My mother's grandfather D. J. Toomey was a builder's apprentice who ran away from his parents' home in Brooklyn in 1871, when he was sixteen. He went first to Chicago, where he had heard that there was work rebuilding after the big fire, then spent five years drifting around the West, looking for "the place." When he was old, someone got him to write down some of his memories of the frontier. There were five stories altogether, each one about five pages long, which my mother typed up and turned into grammatical English.

"My next move [after Chicago] was to Greeley, Colorado," D.J. says in one story:

> Horace Greeley got a concession of 12 miles square on the Platte River 50 miles west of Denver from the Government to establish a college. A Mr. Meek, who was afterward killed by the Ute Indians, lectured all about the beauties of the country. I signed up with many others to take Horace Greeley's advice "Go West Young Man Go

West." In crossing the plain from Kansas City to Denver we passed through buffalo herds. All one day they were so thick one place near Kit Carson and another near Denver they stopped the train when the buffalo stampeded across the track in front of the engine. I did not stay long in Greeley. I went down the Platte and worked on a railroad grade building from Juleberg to Greeley. I got paid off in the spring. By the time the check came I sold it for 20 cents on the dollar. That was the start of the Panic of 1873. I got a job with the government party surveying a Park. Then I got inflammatory rheumatism and spent several months in a Denver hospital. I got out broke and crippled in my feet. I tramped several hundred miles looking for work, I even begged several meals, paying for most of them by hoeing gardens and chopping wood. Finally I got a job moving a farm house into the town of Fort Almans, which led to several jobs at good pay. I got a good gun, a horse and a saddle and went on the buffalo range. I located a ranch right where Sterling is now and killed buffalo and antelope during the winters of 1873 and '74. Then I started for the Black Hills in 1875 from Ogallala, Nebraska. Soldiers were stopping everybody, so I stopped to work punching cows until March 1876, when I started from Sidney, Nebraska. I arrived in Custer on March 27, 1876.

That part of D.J.'s story was a map I had used in my first year on-line. True, I wasn't crippled in the feet, but my computer skills were minimal at first, and this presented me with some problems in my trampings around the Net that my great-grandfather didn't have to worry about. But in the end that map did not take me where I wanted to go. (On the other hand, none of the maps the earliest western pioneers relied on were any good either.) The nineteenth-century pioneer experience in the West, with its myths of self-reliance, independence, and freedom from the restrictive trappings of society, and which was so popular with the cybersalespeople of my day, supplying a kind of *Rand McNally Road Atlas* of metaphors for the cybernaut, was not, in fact, a particularly reliable guide to life on-line. The landscape of the Net was not the great wide-open landscape of buffalo herds and antelope, although I had imagined it was in the first year of my travels. The frontier was more communal now. The frontier lay inside the group.

I. "I Think You're Dead." ◊ ◊ ◊ ◊ ◊ ◊ ◊ ◊ ◊ ◊ ◊ ◊ ◊ ◊ ◊ ◊ ◊ ◊

I rode my bike over to lower Broadway to see my old friend and former neighbor Mark Boyer, in whose pad, in 1987, I had encountered the digital world for the first time. The digital world had been good to Mark. Record-a-Pet International, his first commercial venture in digital technology, had grown into Shoot Digital, a digital-imaging business for commercial photographers. Instead of shooting pictures with regular film and sending that out to be developed at a professional lab with chemicals, the photographers in Mark's place could take pictures with digital cameras, which captured the image in bits; Mark then transferred the bits from those cameras to one of his powerful Macs, brought the images up on one of his screens, and the digital retouchers tweaked them using software like Adobe Photoshop. When they were finished they made a print with one of Mark's digital printers. Mark had recently opened new offices in a building where a lot of fashion shoots took place, and almost immediately had more business than he could handle.

Mark was still a devoted Mac user, and as always he had the latest hardware. We lingered before what Mark had dubbed his "Tower of Power." He had six PowerMac 9500's with 8-gigabyte hard drives, 278 megs of RAM, and "Power PC" microprocessors, which represented the latest iteration of Moore's Law, as well as

Radius Pressview color monitors, scanners, color printers, and assorted other goodies. His hardware gave me the same feeling I got when I went into a really good outdoors store and inspected the ice axes and the crampons.

Mark sat down at one of the machines and took it out for a spin. He brought a photograph up onto the monitor and tweaked the image rapidly, the features stretching, the background changing, imperfections vanishing, all happening at the speed of Mark's flying fingers, while he kept up a steady stream of conversation. He knew how to drive.

When that demonstration was over, we went to the cappuccino bar that was on the same floor as his office (very digital), where models and photographers sometimes lingered between shoots. We talked about the on-line world. Mark lit himself a cigarette with one of his antique Zippo lighters (another one of his enthusiasms). A model came out of one of the studios and sat down at the bar, crossed her long legs, lit a cigarette, and looked at us out of eyes ostensibly slitted against the smoke of the match, but which also conveyed a look of considerable disdain. However, I noticed that as soon as she heard us using words like "bandwidth," "node," and "configure," she began looking at us with much greater interest.

"What do you think of the idea of a virtual community?" I asked Mark. "You know, a group of people who only know each other from on-line. A society of Mind. What do you think of that stuff?"

"I don't really get it," Mark said. "I mean, it sounds good, but let's face it, it's all going to turn out to be one big mall anyway." Mark used the Net mainly for grabbing cool software, like CU-SEEME, a primitive video conferencing system, or for playing games, or for looking up obscure information about popular music. The community aspect of on-line life didn't interest him at all. There seemed to be two basic types of on-line users: those who used it for information and games, and those who used it for community. Mark was the first kind; I was beginning to discover, somewhat to my surprise, that I was the second.

What would it be like to "live" in one of these places? To feel the intimacy of e-mail, combined with the sense of history and social stability that repeated public encounters with a relatively small group of people would create—what would that feel like? Although the amount of "community" one could get through the computer was still severely limited, and it did not come cheap— during my first year in c-space, my total on-line charges came to $1,472, and that did not include my telephone bills—perhaps it was possible to supplement the absence of sound, sight, taste, and touch with rapid and varied, though much shallower, exchanges of "recognition." And as communications power, or "bandwidth," became more plentiful, presumably the cost would go down.

"But what about the theory that if we fill the on-line space with our own culture, it will make it harder for Disney and Time Warner to fill it up with their stuff?"

"I don't buy it," Mark said. "People were spouting that same kind of utopian stuff around the time of the Industrial Revolution, and it didn't include pollution and child labor, right?"

"Well, the Industrial Revolution was built on the backs of workers. Human labor was the weak link in the whole thing. In the Information Revolution we can get machines to do most of the grunt work."

"Maybe. Anyway, I guess the good thing is that cyberspace is infinite," Mark said. "Presumably you could have your utopian community at the same time that Disney has its mall."

That was probably true. But somehow it wasn't as exciting that way. I wanted it to be ours, or theirs—no in-between.

"Hey, you want to go back to the office and play Marathon?" Mark asked.

Marathon was a computer game for Macs that took place inside a spaceship reminiscent of the movie *Alien*. The game had that same intense, trapped-inside quality of the movie. Mark had networked his PowerMacs together so that you could play the game against each other, sitting side by side, and he had really good digital speakers that made the feeling of being in the spaceship very real; you could hear the aliens' tails making scary slithery sounds on the floor as they ran toward you in the darkness.

To demonstrate how Marathon worked, Mark restarted a game he had been playing earlier.

"I think I've already killed a bunch of guys down here," he said, walking around in the dark ship. The foreshortened barrel of his gun, representing his point of view, moved around the bodies and puddles of gore. "Yeah, looks like everybody's dead." He stuck a key into a wall of the spaceship and then ran to get on the elevator going up to the next level, his footsteps making metallic-sounding echoes in the dark space. There was a fearful rustling sound behind him, then the hushed trailing sound of a tail on the floor. The sight of the alien was a letdown compared to the sound. Mark wasted him.

Then Mark and I played against each other, each of us sitting at a computer. I never played games on my computer, and within a few minutes my fingers and wrists were killing me. I had no coordination and kept firing my gun by accident. Mark told me to cut it out because I was wasting ammo. Then he said, "Uh-oh, I can see you on the motion sensor. Actually I'm right behind you." He shot me a couple of times but purposefully didn't kill me, which was decent of him. There were white flashes on my screen and thudding sounds of the bullets whunking into me. I wandered around a little more and then there were gunshots and flashes of red and then I was down and it appeared that Mark was kicking me in the head.

I tried to shoot him, but nothing was happening.

"Um, I think you're dead," Mark said.

On Mark's advice, I set out to get a point-to-point protocol (PPP) connection, by signing up with an Internet service provider (ISP). Such a connection would allow me to dispense with the relentlessly commercial, lowbrow services like CompuServe and AOL, and would give me greater access to other parts of the Internet. But setting up the connection required configuring your computer to meet the requirements of PPP—your computer got its own IP address, making it an official "node" on the Net. To do this I would have to get "closer to the machine," as the digital guys said, than I had ever been before, and just the thought of that filled me with the old anxiety. Then I heard about an "Installer" program that mostly automated the PPP configuration process, available on the Manhattan-based bulletin board ECHO, so I signed up for a PPP account there, downloaded the Installer to my computer, read the Read Me file, and followed the instructions. But I could not get past the second stage of the log-in to my PPP account and was stumped about what I might be doing wrong, so I called one of the technical people at ECHO, whose username was Garbled Uplink, on the telephone. The phone picked up at the other end and I heard an electronic hissing sound, like an on-line sound bleeding over into a voice line, a howling typhoon of bits. Then a voice from within the typhoon said, "Hello?"

"Hello!" I called out. "Yes! Hello! Is this Garbled?"

"Yeah, this is Garbled."

I told Garbled my problem and he diagnosed it as a bad "init" (initialization) string, which is the code the computer sends to the modem to get its attention.

"Try this one—A, N, ampersand which is Shift 7, 1, A," Garbled said. He stayed on the line while my machine went through the log-in stages again.

"It worked!" I said. "I'm in!"

"Congratulations, dude," Garbled said, "you're a node on the Net."

I went down the hall to Dan Henderson's office and said, "Hey, Dan, guess what, I'm a node on the Net!" Dan congratulated me and gave me a floppy disk with a copy of a program called "Fetch" on it, which was a GUI for making on-line file transfers. I went back and installed Fetch on my hard drive, logged on again, and went to a computer at Dartmouth where I found the free programs Telnet and Eudora. I downloaded both—simultaneously!—and installed them. I also downloaded and installed other "freeware," which were the free software programs that various public-spirited programmers had assembled for netizens—Blue Skies (a weather program), JPEGView (a graphics viewer), Newswatcher (a Usenet newsreader), and a variety of screensavers. People who didn't use computers sometimes said that they weren't satisfying because you couldn't break them apart and rebuild them, as you could, say, a lawnmower. What they didn't understand was that you *could* do this, except you did it with software—the gears, rods, and valves of the digital age. You used some software from here, some from there, wrapped it all together with some virtual spit and twine, and you built your own graphical user interface in your living room. And it worked! Before, I had to rely on the interface CompuServe provided, and charged me for; now, for less money than I paid CompuServe, I could use free software that was more useful, powerful, and versatile than anything the Big Three provided its subscribers. With the multi-tasking capability of PPP I could be downloading software in a Fetch window, chatting in a Telnet window, and writing e-mail in a Eudora window—all at the same time. And as was the case with the desktop metaphor, the multi-tasking metaphor soon did a backflip in my head, so that if I was doing two or more things at once—reading, watching TV, and making notes, say—I started thinking of myself as multi-tasking, and before long found myself doing a lot more of it.

II. A Man Smell ◊◊◊◊◊◊◊◊◊◊◊◊◊◊◊◊◊◊◊◊◊◊◊

I lurked in the chat rooms, where people spoke to each other in real time, and which were the most popular public spaces on AOL and CompuServe. They were especially popular among teenagers. My teenage nephew, for example, whose screen name was <Dolf00>, had found a home in a role-playing chat room on AOL. He had met a girl in his chat room named <Melynda>, who, unlike many female users, actually gave <Melynda> as her username. Whenever <Dolf00> logged onto AOL—"Welcome!"—he would use the "member on-line" function to see if <Melynda> was signed on. If she was, they would jointly open a private chat window on their screens and type messages to each other, while watching the group discussion scroll past in another window. It was like meeting a friend at your locker and whispering to each other while the other kids passed in the hall. While they were chatting, other people in the public window would sometimes send <Melynda> rude messages like, "Hey, Melynda, want to play my flute?," because <Melynda> said in her member profile that she is a flutist.

One day <Dolf00> told his mother that <Melynda> would be visiting for the July Fourth weekend. She was spending her own money to fly from Cincinnati. Her mother and my sister talked on the phone beforehand. My sister is a Wall Street lawyer; <Melynda>'s mother said she was a single mom who worked part time in a store. It seemed like a perfect example of the ability of the Net to bring together people from different worlds.

But such encounters don't remain idealized for long in the real world. For one thing, <Melynda> turned out to be pretty fat. She was the kind of firm fat where your legs are slim and you look pretty good, but she was still pretty darn fat. She was also very quiet and shy; even <Dolf00> thought she was shy, insofar as I could penetrate his inscrutable mind. They stayed in <Dolf00>'s room a lot with the door closed, going on-line to the chat room, although it didn't seem to be as fun when they were actually together.

Not long after that <Melynda> disappeared from AOL. <Dolf00> would only say, "Melynda logged out." Did he care? Hard to tell. Before long, he had met another pal, a twenty-three-year-old house-wife whose husband, she claimed, talked to her for about twenty minutes a week. <Dolf00> was a better listener. In fact, he was all ears.

Most systems kept no records of your conversations in Chat (you hoped), so people were more inclined to say things they might later regret having said than the posters in forums, newsgroups, and on BBS's, where a record of your remarks was preserved, at least for a week or two. The relative anonymity of Chat was a big part of its attraction. Identity was fluid and accountability was low. Chat was the cyberforest of Arden. The weirdest Chat story that I ever heard was sent to me by a stranger on e-mail. I sent him mail back asking if it really was true, and he swore it was.

When my sister attended the U of M/Duluth while I was here, at the U of M in Minneapolis we wrote via e-mail nearly every day. Then we started meeting each other in chat rooms. She was INNO94, I was MACOMBR. One time, she changed her name and spoke to me anonymously, which was stupid and inconsider-ate. (She's 21, four years younger than I am). Our conversation sort of escalated. We moved from the chat room to her personal message box and talked for about an hour.

Without knowing it, he committed cyberincest with his sister.

She ended up disappearing from the chat room, abandoning our conversation, and told me the next day via e-mail that she had been MARGO—the woman in the chat room. I was supposed to catch on to her name, she said, because there's an inside joke with the names. But I hadn't caught on. I had said a lot of things to her I wish I hadn't. So Thanksgivings and other holidays are awkward.

There was a lot of sex on Chat. Unlike talking to a stranger on a telephone, where the sound of another person's voice might activate certain taboos about random, promiscuous sex, in Chat only written words passed between people, a level of discourse at which the superego apparently did not operate for many users. And sex was, after all, one of the great ways of enjoying the expanded personal freedom that computer networks offered. Whether what people were doing constituted real sex, or whether it was a kind of phone sex, or whether it was just high-tech masturbation, was an interesting question, although thinking for too long about the question could make you develop warts on your hands. A lot of cybersex, it seemed to me, wasn't even about sex at all. Sex was merely the easiest and most obvious form of discourse for two people who knew nothing about each other, and perhaps had nothing to say to each other, to borrow in order to have a half-interesting conversation. What a lot of Chat in general, and cybersex in particular, seemed to be about was simply the desire to be *on the line*—to take the flow of bits that another user sent you, add your own flow to it, and see where it went. Chat was like playing that kids' game in which you stretch your palms out, and the other person stretches his hands out on top of your palms, and you try to slap the back of the other person's hands. There was something aggressive about it, but there was something tender about it, too, when you were in sync, when you were feinting with your wrist muscles, up and down, not even intending to slap, although the other person doesn't know that, so that his nerves twitch with your nerves, each of you superaware of each nuance in your fingers and wrists, as if your nerves are twined together.

When I first encountered Chat, I was a little taken aback by my impulse to try cybersex, and made an effort to put the thought out of my head, but it would not go away. So one evening I logged onto the CB Simulator, which was CompuServe's Chat area, as <Bambi>. I entered a chat room. Everyone in the room at that time saw the words "Bambi is Here" flash on their screens. Within thirty seconds, four male-sounding handles had sent me private messages—the sort of complimentary, somewhat fresh little nothings with which a man can turn a pretty girl's head—each inviting me to join him in a "private room." I accepted the offer of a gentleman named <candyman>. Then it occurred to me that whatever I was about to do might some day constitute adultery. It didn't count yet, because the moral universe of bits was not fully resolved. But if this practice became adultery in the future, whatever I was about to do now might be grandfathered in.

I called out to Lisa, who was in the next room, "Hey, I think I'm about to get laid here for the first time on the screen!"

"WHAT?" came loudly through the Sheetrock. Lisa appeared in my doorway.

"I mean, as a woman," I said.

"You're a woman?"

"I'm a woman who may also be a deer. My name is Bambi. I'm sorry you had to find out this way."

> candyman: Would you like to run through the woods
>   together?

I explained about the CB Simulator and then pointed at my screen and said, "So you see this gentleman, the candyman, has invited me to join him in a private room."

"And you accepted his invitation?"

"Well, I thought he was sweet. Now he's turning out to be a wolf."

Lisa leaned over my shoulder—our second moment of togetherness in front of my screen—and peered silently at the crawling sentences for a while.

"So do you?" she whispered into my ear, in a sexy Bambi-like voice.

"Do I what?"

"Do you want to run through the woods with the . . . mmmmmm . . . *the candyman?*"

Hmmmm. An interestingly modern situation. Yes, in fact, I did want to run through the woods with the candyman. The thought gave me a funny tingling feeling I had never felt before. I covertly studied Lisa's profile but did not see the hardening in the jaw muscle that is the thing you don't want to see, ever, if you can help it. Could it be that, thanks to the miracle of the many-to-many, we were about to get into a *ménage à trois?* And with two women? O, utopia! Although I had never imagined that I would be one of the women. . . .

"A little run through the woods might be fun," I suggested, tentatively. "I am a deer, after all."

I typed:

It's dangerous. There's a man smell.

The candyman typed back:

Oh, I mean no harm.
Bambi: But you might eat me.
candyman: Would that be so bad?
Bambi: It might hurt.
candyman: Are you in season?
Bambi: Are you rutting?
candyman: Maybe. Could I take a nibble?
Bambi: OK

"Okay?" Lisa said. "You're going to let him eat you?"

"You don't want me to?"

> candyman: Where can I start? You tell me.
> Bambi: On my tits.
> candyman: A great place to start.

Once again, I quickly cut my eyes away from screen, toward Lisa's jawbone, but still saw no sign of serious trouble.

> candyman: Can I gently suck on them?
> Bambi: Well, lick around them first.
> candyman: OK, are your nipples hard?
> Bambi: All six of them.
> candyman: Six!! Can I handle them all?
> Bambi: You may.

" 'You may'! How ladylike!"

> candyman: Would you mind if I started to work a little
>   lower?

Somehow I knew this was coming. The giant electronic brain that was the Internet seemed to have a one-track mind! I wasn't sure I was ready for this, not now anyway, and, besides, in my imagination my digital pudendum was starting to morph into something between a woman's and a deer's, which I didn't even want to be thinking about.

> Bambi: Well, you might be surprised what you find down
>   there.
> candyman: What is that? Can you tell me?

Bambi: It's bigger than a breadbox.

candyman: Oh no, don't tell me you're a man. I only prefer fawns.

Bambi: I understand. Nice chatting with you.

candyman: But, but your handle—sounds like a female.

Bambi: It's actually my real name. My mother named me Bambi and I have been living it down for the rest of my life.

candyman: Oh, sorry.

Bambi: Why do you call yourself candyman?

candyman: I usually give candy to the women in the office where I work.

Bambi: In return for sexual favors? Where do you work?

candyman: In a hospital.

Bambi: Are you a doctor?

candyman: Bambi, are you a male or female? I'm still confused.

Bambi: You're not the only one.

"That makes three of us," Lisa said.

candyman: No, I'm not a doctor, but would you like to play doctor?

Bambi: I'd like to be the doctor if you'll be the nurse.

"What if I make candyman the girl? Would that be better?" I asked, luxuriating in the pleasure of the extra split-second that you got to think before answering, with no pressure on you, which was part of the pleasure of Chat. Speechwise, I am like the figure skater who will land the triple axel eight times in a row in practice and then miss it in competition. In the on-line medium you could always land the triple axel.

"I'm not sure," Lisa said. "I think that might be worse."

> candyman: Ah, this is starting to sound a little strange for me.
> Bambi: Oh Candy?
> candyman: Yes Bambi.
> Bambi: Scalpel please.
> candyman: Oh no, it's Lorena.
> Bambi: But I can sew it back on too.
> candyman: Aaugh!
> SYSTEM: candyman has left CB.

"I don't think that turned the candyman on," Lisa said.

Later, when Lisa wasn't around, I returned to the chat rooms (this time as <Debbi>) and allowed a very forceful gentleman named <BlackPrince> to go all the way with me, until finally, after what seemed like a lot of work on my part, I saw the words

> BlackPrince: Aaaaaarrrhhh I'mmmmcommmmminnnnnggg

appear on my screen, and I dutifully typed back

> Debbi: Aaaaaaarrrhhhi'mmmmmcummmmmiinnnnnnggggg
> tooooooooooooooo

—instinctively acquiring, I guess, the art of faking it. The following day I received an e-mail message from <BlackPrince> inviting me to an f2f party in a bar in Buffalo, New York.

Oh, so maybe that was it: cybersex was a way of recruiting people into real sex. That at least made sense to me. Masturbating while typing sexy remarks back and forth did not make sense to me (technically *or* emotionally).

Chat was also a useful tool for making a nonsexual IRL relationship sexual. You'd meet a stranger at a party, or in an office-supply store, start a conversation about the Internet—because the Net was as good an icebreaker as any these days—and exchange e-mail addresses. On-line, you'd arrange to meet in one of the chat rooms, and there you'd start chatting in a much less inhibited manner than you would face-to-face or on the telephone. Chat was thus like a pander, a go-between. Talking on the phone was the next step—but that was a mighty step. In Chat it seemed unreal, a bit as though it was all only happening in your imagination. But each time you went back to the chat room to meet your <BlackPrince> you made it more real, until finally you had to take it to the next level— the phone, usually—or break off the relationship altogether. These were the horns of the dilemma on which many an infonaut was caught.

III. The Groupmind ◊

Asynchronous postings, which were like public e-mail—the kind of communication conducted on Usenet, the mailing lists, the Compu-Serve forums, and the bulletin boards—were generally less anony-mous than Chat. Because a record of your conversations was more likely to remain in existence, people tended to be more cautious about what they said. As more conversation was added to the record over time, it acquired a history; then the history was interpreted into a kind of ideology. The record not only existed, it stood for some-thing. It was the history of all the transactions between the people who belonged to that particular group.

Bulletin boards were apt to be the most accountable places in cyberspace, because the number of users was relatively small and stable, and the postings were preserved for years. You could usually get out onto the Net from a bulletin board, but people from out on the Net could not get in without a subscription. ECHO was my neighborhood bulletin board, as well as my ISP, so naturally it seemed like a logical place to look for a home. Its first physical location, which was in the Greenwich Village apartment of the owner of the system, Stacy Horn, was only a five-minute bike ride from our place. She had fifty-two telephone lines running into the apartment (for a while no one else on her street could get a new line), connected to fifty-two modems, which had red lights on them that constantly blinked with incoming calls; the modems were arranged on shelves at one end of her railroad apartment, like an art installation. Christmas lights were strung around the top of the room. Horn was a petite woman in her mid-thirties, with thick black hair that framed her face like Cleopatra. She was born with a hole in her heart so she had to stay inside a lot as a child, and this, Horn believed, began her interest in on-line communication. "Communicating without people being right there . . . where we are, who we are . . . people who want to communicate no matter what—that's what inspired me," she told me. Her core population on ECHO were downtown Manhattan artists, writers, performers, and hangers around.

For a while, Horn's chief technician was a twenty-two-year-old digital guy named Phiber Optik. Optik, as he was known to his friends, was a famous "cracker." Crackers were loosely allied with hackers under the "Information Wants to Be Free" banner, but there were important differences between the two cultures. The classic hacker was a hippie-nerd who had flourished in the age when the personal computer was a disconnected box. They were mostly electronics wizards who cobbled together bits of electronics and software to make primitive computers. Some were inspired by ideas of social change and citizen empowerment, which they were helping to promote by putting computers into the hands of ordinary people. The rise of the Net had bred a new generation, the crackers, whose genius was not building computers and networks, but breaking into them. Optik, for example, was said to be able to visualize network architecture in his head as he was probing a new system, as though he were playing a video game, moving up and down through levels. Optik's public image was that of a "good" cracker, a lad who cracked for the pure joy of sneaking into, say, an AT&T system and leaving behind messages that would upset a bunch of guys in suits. Whereas his peer, the infamous Kevin Mitnick, cracked with malicious intent. Actually, although Mitnick had tried to do lots of damage, he did relatively little, but Mitnick was reviled as a pariah, while Optik was celebrated as a swashbuckling hero and a martyr to the cause of free information.

Optik was small and thin and dark, smoked furiously, and rarely smiled. His sleeping habits were famously irregular. "Phiber had to go home and get some sleep," someone would say when you called Horn's apartment in the middle of the day to report a technical problem. Reporters and fans often stopped by Stacy's place to pay homage to Optik, which would make him even more sullen. "Getting compliments makes me angry," he would say. He was a media darling.

I also signed up for another bulletin board called the WELL, which was based on the other side of the country, in the San Francisco Bay area. The WELL was older than ECHO, larger, and its history was more rooted in the hippie romanticism of the 1960s, not in the club and performance art scene of Manhattan. The WELL had grown out of the 1970s back-to-the-land-through-technology idealism embodied by the *Whole Earth Catalog*. The basic idea was that by providing citizens with the technology to do more things for themselves—to grow their own food, make their own clothes, build their own wells, design their own solar-heating systems, and, now, make their own media—you could free people from their dependence on mass consumer products and corporate marketing, which were the windows through which the soul leaked out of modern man. The WELL was a digital version of that idea, an attempt to wed the enthusiasm for computer technology on the one hand to a romantic, pastoral ideal of life on the other, and in doing so to solve a basic contradiction in American society that could be traced back to the thought of Thomas Jefferson. How could Jefferson not have foreseen that the mechanical inventions he loved, like the steam engine, would bring the factories that he deplored to the American landscape, and then the railroads, and then the freeways and the malls, and spell the end of his beloved agrarian husbandry? Or could Jefferson see even further into the future, to a time when machines would restore the pastoral to our modern mechanized age?

Shortly after signing up on the WELL I read a book called *The Virtual Community,* written by Howard Rheingold, a former editor of the *Whole Earth Review,* who was a longtime member of the WELL, where he was <hlr>. In his book Rheingold wrote that when he first encountered the WELL, he said to himself, "This is groupmind." The basic idea of groupmind was that the network was a kind of giant brain that was made of all the people's brains on the network, wired together. It was like my facile idea of the giant electronic brain that would answer all my questions, except the brain was made of people, not machinery. Rheingold seemed to have come to his idea of groupmind from the East, from Zen Buddhism, whereas I was approaching from the West, from the German Romantic idea of the "Over-Soul," popularized in England by Coleridge and Carlyle, and then in America by Emerson (who read widely in eastern literature). In his essay "The Over-Soul" he wrote:

> In all conversation between two persons, tacit reference is made, as to a third party, to a common nature. . . . And so in groups where

debate is earnest, and especially on high questions, the company becomes aware that the thought rises to an equal level in all bosoms, that all have a spiritual property in what is said, as well as the sayer. They all become wiser than they were. It arches over them, like a temple, this unity of thought, in which every heart beats with nobler sense of power and duty, and thinks and acts with unusual solemnity.

Rheingold was suggesting the existence of something like the Over-Soul on the WELL.

The sensation of personally participating in an ongoing process of group problem-solving—whether the problem is a tick on my daughter's head or an opportunity to help policy-makers build a public network—electrified me. The feeling of tapping into this multi-brained organism of collective expertise reminds me of the conversion experience the ARPA pioneers describe when they recall their first encounters with interactive computers.

That sounded like what I was looking for, and I found myself coming back to the WELL.

A distinctive feature of the WELL was its conferencing software, a relatively difficult-to-learn keyboard command-based interface known as "Picospan," which had been written by a programmer named Marcus Watts. There was no GUI: to navigate the WELL you needed to learn the keyboard commands—words like "go," "browse," and "see since -5." Another aspect of the local culture, which marked the WELL indelibly, was that it was impossible for users to be anonymous. You could change your "pseud," which was the line that appeared next to your username, as often as you wished, but you could not change your username without starting a new account. And by typing "bio" and then the username, you could find out the user's real name. Although this restricted the freedom of individuals somewhat—because anonymity was, for better or worse, one of the most powerful forms of freedom that existed on-line—it gave a measure of personal accountability to the discussions that was missing from some of the newsgroups on Usenet (where complete anonymity *was* possible, thanks to a guy in Finland who had set up a computer that stripped the identity off messages before sending them on their way). Partial anonymity was also possible on ECHO, America Online, and CompuServe.

One reason for the WELL's policy was the founders' knowledge of an earlier conferencing network called EIES, established in 1982 by a group of forty people associated with a research institute in La Jolla, California. For about six months, the participants were caught up in the special lightness of hope and possibility that the technology could inspire, until one day a member of the group began provoking the others with anonymous on-line taunts. Before long, the community was so absorbed in trying to identify the bad apple that constructive discourse ceased. The groupmind posted many messages pleading with the individual who was doing this to stop, but the poster didn't stop, and groupmind was destroyed. And not only did this break up the on-line community—it permanently affected the trust that those people had for each other in the face-to-face world, because they were never able to figure out who did it. To this day, they don't know.

The total number of people who subscribed to the WELL in 1994 was about ten thousand. Anyone could join. Subscribers paid fifteen dollars a month, plus two dollars for every hour they were on the system, plus their telephone charges—it was easy to spend a hundred dollars a month. The average user was white, male, and well off, although women were a stronger presence on the WELL than on the Net at large. The WELL had studied its user population and concluded that the lurkers on the system outnumbered the posters by about nine to one. (A similar phenomenon could be observed on Usenet, where most of the postings were made by 2–4 percent of the population.) The lurkers were either not using the community at all, or they were using it like television, as something to watch. On the other hand, people who *did* post tended to post often. In the course of a day an active poster might compose long, well-written posts in five or six different topics.

At first I merely lurked on the WELL. I was a username, which was <seabrook>, and that was all. (Some of the other users chose names like <rabar> and <humdog>, but that was too far out there for me, so I went with <seabrook>, which in some ways seemed a better username than a surname anyway; as a surname it sounded a little too much like an adult retirement community or a nuclear power plant.) Aristotle said that action determines character, and if he had been on-line he might have said that posting determines character. As a lurker, my existence on the WELL was the ontological equivalent of a tree falling in the forest with no one to hear it.

I mainly lurked in the News, Media, and Books "Conferences," picking up bits of local slang, custom, and lore. I learned that members of the WELL community were sometimes called "pern," a slightly sardonic genderless pronoun invented to replace the politically problematic "he/she" formulation. Some pern took offense if you presumed to refer to them by their real names, but didn't know them IRL. God was called "gopod" on the WELL, because a pern named <jrc>, who was IRL Jon Carroll, a columnist for the *San Francisco Chronicle,* had once committed that typo while typing the word "god," and the word "gopod" was spread across the system by a phenomenon known as a "meme." A meme was a concept made popular by the biologist and writer Richard Dawkins, who defined it as a unit of cultural transmission, or a unit of imitation. . . . Examples of memes are tunes, ideas, catch-phrases, clothes, fashions, ways of making pots or of building arches. Just as genes propagate themselves in the gene pool by leaping from body to body via sperm and eggs, so memes propagate themselves in the meme pool by leaping from brain to brain via a process which, in the broad sense, can be called imitation.

Computer networks in general were hotbeds of jokes, myths, gossip, slang, urban legends, and many other forms of memes. The concept of the "meme" was itself a meme on the WELL.

Most of the time the WELL was peaceful and bucolic, just a bunch of intelligent and articulate "folks" talking about kids, books, travel, the San Francisco Giants, Buddhism, the Dead, current events, computers, the state of the WELL, and many, many other subjects. But every now and then a thread would erupt into what was known on the WELL as a "thrash." Although thrashes always ended up being about large issues like anonymity, tolerance, copyright, privacy, or community-management relations, they often started small. Someone would throw a stone into the conversation, mainly to see the ripple it made, then someone else would throw a bigger stone, then a third pern would find herself splashed by these stone-throwing macaques and splash back (to borrow a metaphor favored by a pern named <tigereye>, whose postings were a wild gleam of intelligence in the jungle). Then one of the aging silverbacks in the tribe would take it upon himself to teach these impudent youngsters a lesson. Suddenly, people you had been chatting with civilly for weeks or months were locked into a wild, raging battle. There was so much emotion involved! Impassioned posters leapt to the defense of the stone-throwing macaques, others posted ad hominem attacks on these defenders; troublemakers provoked, peacemakers called for calm, oldpern wearily referred to past thrashes; but most of all everyone earnestly tried to get at the *truth* of whatever that particular thrash was about, to sum it up in a single posting, to make the speech act that would set everything right again. Then the thrash would begin to gutter and quickly die down and deep silence would settle over the jungle again. The antagonists would retire to their separate corners, and the rest of the tribe would go back to grooming one another until the next thrash erupted.

IV. A Natural History Museum of the Sixties ◊ ◊ ◊ ◊ ◊ ◊ ◊ ◊

I spent time reading in the Archives conference, where the history of the WELL was stored. The odd thing was, when you were in Archives, you really did feel like you were in the presence of history. It was the history of all the time all the people on the WELL had spent writing to each other—so much thought, emotion, effort, time, spite, and goodwill was stored there. Browsing in Archives reminded me of walking through the Princeton boathouse in the dim light after crew practice, with those big lacquered wooden rowing shells bulking out at you in the gleaming dimness. Except that in Archives the shells were made of words.

In Archives one could learn that the WELL had been cofounded by Stewart Brand, the writer and media philosopher. Brand had once been a member of the Merry Pranksters—Ken Kesey's troupe of hippie performance artists who rode around the country in a magic bus driven by Neal Cassady, the model for Dean Moriarty in Jack Kerouac's novel *On the Road,* and who presided over the Acid Tests, a series of LSD-inspired be-ins. Out of the Acid Tests came the Trips Festival, a multimedia extravaganza that was the prototype of the modern rock concert. In the mid-1970s Brand put on the Hacker's Conference, an important early intersection of visionary computer programmers that would flower in the Net culture of the 1980s.

In the early 1980s Brand began to see in computer networks a way of preserving some of the goals of the Acid Tests and the Trips Festival: lights, sound, multimedia, and a group of minds that could be connected in the way that sharing acid connected you, but without the drugs. (As Brand was fond of saying, the problem with LSD was that it never got any better.) Together with another 1960s visionary named Larry Brilliant, who had a company called NETI, Brand founded the WELL in 1984. Brilliant licensed Picospan to Brand in return for half the company. The system ran on a VAX microcomputer in the *Whole Earth Review*'s office, which was on Gate 5 Road in Sausalito.

Here in Archives, preserved in the amber of words, was the counterculture that I had been too young and too square to participate in the first time around. There were bit-mapped versions of former communards from Morningstar Ranch, Diggers, Brightonians, cowboy Zen folks, and Buddhists from the Esalen Institute. One of the WELL's early members, now one of the elders of the community, was a radio host named David Gans, whose first incarnation was <maddog>, and who later became <tnf>, and who had brought a group of Grateful Dead fans to the network. A computer network was an ideal way for devoted Deadheads to stay informed of concerts, organize get-togethers, post set lists, and debate whether or not the version of "Truckin' " that the band played last night in Philly was as good as the version they played in Denver four years ago. Above all, the WELL was a place for Deadheads to meet when the Dead was between gigs.

Three early, important members of the WELL had come from a famous commune, The Farm, which was started by a charismatic leader named Stephen Gaskin. In 1971 Gaskin had led a caravan of psychedelic buses across the country, recruiting more longhairs along the way—"a strange organism moving through the blood-stream of America," as one Farm member later put it on the WELL. Finally Gaskin stopped the buses in the middle of a field in rural Tennessee, and that became The Farm. Over time Gaskin began to display megalomaniacal tendencies, and by the early 1980s many of the original Farm members had left the commune. Brand recruited a former communard named Matthew McClure to be employee/innkeeper on the WELL, and McClure, who was <mm>, recruited two more Farm veterans, Clifford Figallo, who was <fig>, and John Coate, who was <tex>. Among the lessons they brought to the WELL, Brand told me on e-mail, were:

> Don't overwork and underappreciate the females, or they leave, and then the party is over. Don't invest much in a charismatic leader: he will steal everything you've got and then blame you for not having any more. It takes more than a sharp stick and earnestness to make a garden produce food. Don't piss off the neighbors. Stuff like that.

When I ran Brand's remarks by one of the females on the WELL, she responded, "This is so fucking condescending. Such bullshit"—which made me think that perhaps things hadn't changed that much since The Farm.

Stored in Archives was conversation that <fig>, <tex>, <mm>, and a few others had had on the system in 1987 about life on the Farm.

Topic 44 [archives]: Communes—Past, Present, and Pluperfect
#18: Clifford Figallo (fig)

We were very much into Truth, and at times we wielded it like a bludgeon. It was for their own good, of course, but it felt *so good* to lay a big fat TRUTH on someone. Almost as good as it felt bad to have one laid on *you*. Some folks (we called people "folks", folks) were "tennis ball eaters." You would serve up your best, most compassionately worded explanation about how and why they were assholes, and for a return you would get some lame reply that sounded like they deliberately missed the point. Grrr, that was frustrating. Especially when you were staying up until 2AM just to "get into their thing"; and it was for their own good!

<tex> started talking about his experiences with Stephen Gaskin.

#25: John Coate (tex)

As for me, my relationship with him deteriorated within the first year of my being there. Soon after my arrival I formed an R&R band and we got work around the nearby towns. He got the idea that we weren't presenting enough of a spiritual, high musical experience (although we never did play "Whipping Post") and one day when I bumped into him out in front of the main house he said to me in a very loud voice that was easily heard by the 25 or so people close by, "I'm taking you out of the band. Turn in your guitar. You think you're this nice guy, but you ain't a nice guy. You're Jack the Ripper. You're a (now yelling) SON OF A BITCH!!! You have the morals of a snake. Now get out of here!"

Tough stuff to hear, and I was so intimidated that I couldn't come up with the words to defend myself.

#40: Paul Hoffman (phoffman)

Jeez, this Farm stuff sure sounds cruddy to me. These stories remind me of folks who feel that getting the shit kicked out of you in the Marines (or the Catholic church or on a football team) makes you a better person. The Farm, from my meager reading here, sounds like the head-trip equivalent of the Marines. . . .

It probably was like the Marines. Stephen had been a Marine in the Korean War, and he repeatedly maintained that he wanted our gate run like the gate at Camp Pendleton. Very few of us has been in the service, so the comparison stopped there for me.

As it apparently had been for these oldpern, the commune became an early, controlling metaphor for my understanding community life on the WELL. (On ECHO, the controlling metaphor was the trendy Manhattan nightclub, which did not click for me in the same way.) In my mind, the WELL was a cross between The Farm and Brook Farm, the transcendentalist community that was established in West Roxbury, Massachusetts, in the 1840s, which its founders envisioned as "a society of liberal, intelligent and cultivated persons, whose relations with each other would permit a more wholesome and simple life than can be led amidst the pressures of our competitive institutions."

Life on The Farm did sound a little intimidating at times, but on the other hand, the idea that there might be a digital version of Stephen Gaskin lurking somewhere around the WELL, who was going to take a crack at getting into *my* thing, was actually sort of attractive to me. As a privileged white kid who had had his groups chosen carefully by his father, and had learned to choose them carefully himself, I was probably just the sort of person with whom Gaskin could have had a field day when he was in his prime.

In 1992, with the number of users growing rapidly, the WELL began to have serious problems handling the load. The frustration of not being able to log on to your virtual community is like the frustration of being stuck in traffic when you want to be home, but it has an especially annoying quality all its own because you are paying money for the service. When you finally do get in, you are liable to be one pretty frustrated pern. Someone had started a topic in the News Conference to complain about the system problems, and certain posters in that thread had begun to attack Stewart Brand personally, who was on the WELL board, for what they perceived to be his arrogance and lack of concern for other pern, and a thrash had ensued. Brand had then opened another topic to talk about that assumption.

Topic 173 [archives]: System Scapegoat—does the WELL always need one?

Started by: Stewart Brand (sbb)

This is a topic for public discussion of my shortcomings and how they are the source of all problems on the WELL.

The purpose is to determine what is the proper behavior toward whoever is the current System Scapegoat, and how the goat should best behave.

#15: well's cargo (dlee)

Maybe it's because I don't frequent the "right" conferences, but it seems pretty rare to see an sbb posting. I have great respect for his accomplishments, and like Howard says, it seems fairly natural to be both fascinated and intimidated by Stewart. But Howard's description of him as "aloof" and paternal fits what little I've see of his postings, and it seems like Stewart's happy enough being aloof and intimidating.

It fits the pattern that he would open this topic, and with the words he used. I didn't see that, by WELL standards, he's been flamed that much, but he seems mightily annoyed that people would be so low class as to question him at all. Scapegoat? Hardly, Stewart; you flatter yourself.

If you're content to rest on your laurels, then fine, Stewart. Go away. That's not what the WELL's about. We're all peers here, dude. If you think you're being treated unfairly then tell us why, and tell us what IS going on. Give us something more than "wise" terseness. If you really care about us, and the ongoing health and viability of the WELL, then be a part of it and us. Otherwise it's easy to think that you see the WELL more as another nice item for your biography.

#36: Stewart Brand (sbb)

Who actually does the hard work of scapegoating?

Judging by mine, they are volunteers, not paid, self-appointed, not elected or otherwise appointed, and energetic. Unrelentingness is part of the profile. If the scapegoat's behavior is not changing as demanded, hit harder and more often.

Sometimes a Tormenter goes so over the top that he or she flips into the Scapegoat role, for over-reaching and thus endangering the system. Gans has been there, I believe.

Tormenters never get to be the System Darling, who is the mirror of the System Scapegoat. I've been the Darling a time or two. Last time I checked Cliff Figallo was sort of it, but John Coate is carrying the main burden, this week. Darlinghood is limited because you're only allowed to say adorable things online, same as the Goat is only allowed to say defensive things on-line (if he or she says something adorable, it must be attacked).

I've learned there are only two viable responses to Scapegoathood—defensiveness, and defiance. This topic represents my changing gears from the one to other. That I'm having a good time with the experiment is driving my Tormenters crazy, as it is intended to.

#43: Stewart Brand (sbb)

Now a tricky moral question.

Do Scapegoats feel pain?

Of course not. A Scapegoat is a chimaeric entity, a projection, and therefore incapable of pain.

Well then, do Tormenters intend to inflict pain?

Yes, and it apparently gives them pleasure to do so, but that may be my biased perspective. However, they know that a chimaeric entity feels no pain after all, so they are absolved from feeling bad.

This is a demonstration of a fundamental truth, quickly and harshly learned, of Scapegoathood. Never show blood in a flock of chickens.

#57: Mouthy Scorp (axon)

well, heaven forfend that i should bully saint stewart. i'm afraid this latest exercise in obfuscation fails to sustain my interest. it's just a lot of handwaving.

now that i've been branded a Torturer i suppose i can die happy. but it isn't true that my intention is to inflict pain. far from it. my only intention is to help.

i'm not trying to bully you. i'm trying to determine if there's any competence or even concern among the leadership. sorry if you took it wrong, and my apologies to all if my rhetoric offends. but it seems to have gotten your attention, at least, and i'm willing to be "the bad guy" (new well cultural icon; alongside The Scapegoat, The Tormentor and The Darling) if it will give you a safe way to gauge the depth of user resentment without sacrificing your own ego needs.

After scanning this topic as it scrolled in at 2400, I'll merely observe that I regard Stewart as largely irrelevant to the daily workings of the WELL and this topic as a colossal ego-trip.

Around the time of that thrash, half of the WELL was bought from the Whole Earth organization by a businessman named Bruce Katz; in 1994 he bought the other half. Brand stepped down from the board and <sbb> cut back on his use of the system. After that, Brand told me later on e-mail, "I never loved the WELL again, nor fully trusted its process."

V. The Birth of Point-and-Click Thinking ◊ ◊ ◊ ◊ ◊ ◊ ◊ ◊ ◊ ◊

Meanwhile, as I was telnetting around the Net in search of a place to call home, posting here and there but not yet putting down any roots, the World Wide Web arrived. The Web, which in the space of roughly a year would transform the culture of the on-line world from something resembling an engineers' common room into ESPN2, was invented by a British computer scientist named Tim Berners-Lee, who first posted news of it in 1991, in the Usenet newsgroup alt.hypertext. I began to hear of its wonders in mid-1994. At first I tried to explore it using a text-based "browser" called Lynx, but it was not until I downloaded the browser made by Netscape that the real advantages of the Web struck me. Previously the Net had been a little too hard to use. It required a level of software knowledge that most people did not possess, or care to learn. To move around among the computers, you had to know their names, or their IP addresses, and then laboriously type them into your FTP software. But the Web plus the Netscape browser created a much friendlier, easier, point-and-click type of operating environment— basically, the Web made it possible to use the Net with a mouse. The Web also made it easier to view graphics and play sounds on-line. Instead of downloading the images and sounds from the Net to your computer and then using another program to decode them, you could "play" pictures and sounds in your computer's memory, using the browser. The Web was the five-hundred-channel universe that the cable TV people like John Malone of TCI had been talking about for several years, except that the channels were called "sites," and there were five million of them.

There was no forward or backward on the Web, no row of numbers to give the user the feeling of order, like you get from TV. I am a fairly accomplished television surfer, but surfing a TV was like surfing Long Island, compared with surfing the Web's Waimea Bay. In my channel surfing I would typically browse quickly through the first forty-six channels, from New York 1 to Bravo, target three or four good things to watch—a basketball game, a good old movie I'd seen before, MTV, and maybe a nature show on the Discovery Channel—and then click back and forth between these channels at commercials or slow spots. Surfing the Web turned out to be a different enterprise altogether. Although the mouse appeared to be inferior to the remote control, as a surfing tool it actually had far more control over what was on the screen. A remote control changed the channels but had no effect on the programming, so what you weren't watching you missed. On the Web, the programming was designed with the technology of the mouse in mind, so the "show" never began until you arrived. Also, the flow of information across the computer screen was different from the flow on the TV. The Web flow was composed of text, images, and sounds, all jumbled together—weather maps, pictures of Cindy Crawford, the Pat Buchanan for President Home Page, the fake Pat Buchanan page, which had swastikas in place of the American flags, the March Playmate of the Month, pictures of Sharon Stone and Kate Moss topless, exotic cannabis seed catalogues, ski reports from Sugarbush Mountain, and the sounds of Socks, the President's cat—all bobbing in a kind of thick crust, like an icepack floating on a partly frozen river.

The art to Web surfing seemed to be knowing how to skate like a waterbug across the surface tension of Webalicious sites, never lingering long enough in any one place to break through the crust, borne forward by the pursuit of media satisfaction, which always seemed to lie tantalizingly just beyond your feelers. The trick was to hit a site, browse it, see a link, click on it, and get transferred to another site, but then, if the "chug factor" seemed high (the excruciatingly long time an image took to load in RAM), to go up to the menus, click on Go, and drag down to an earlier page, or surf ahead to another site until the previous image loaded.

I didn't like the Web at first. Yes, in theory, surfing the Web was more mindful than watching MTV, because you were pulling content toward you, instead of having it pushed. My brain was having to expend more energy, in making choices about what I wanted to see. But I still felt bored—I wasn't a couch potato; I was a deskchair cauliflower. Maybe this feeling was mainly the result of the chug factor, and would go away when the TV and the computer fully converged, or maybe the boredom was a portal through which the secret path to nirvana lay. It was a particular kind of boredom caused by the bombardment of electronic information, a kind of e-ennui. It seemed as though the point-and-click technology had so permeated the content on the Web that it had created a kind of point-and-click thinking. Sometimes I wanted to say to my screen Yo, I have a *head* on my shoulders, not a mouse! A finger is not poised over my brain, waiting to click it, if it gets the least bit bored! At least not that I am aware of! I don't need little sentencelets and thoughtlets and factlets in order to keep my mindlet happy!

VI. **Mr. E-mail** ◊◊◊◊◊◊◊◊◊◊◊◊◊◊◊◊◊◊◊◊◊◊◊◊◊

When I wasn't in front of the screen, I spent a fair amount of time with my "print friends"—the writers and editors I knew from work. Since Lisa was also originally a "print friend," we had a lot of friends in the medium. Most of the writers I knew were like me—nontechnical types who had started out using typewriters, and in recent years had reluctantly switched to computers and word processing programs, about which they often complained bitterly. Almost none of my print friends were on-line, and most had no plans to be. Many were suspicious of my enthusiasm, perhaps suspecting me of having gone native, or at least of having tapped into some geeky weirdness that I had been doing a reasonably good job of concealing up until the time I discovered the on-line world.

We sometimes ran into each other at publishing parties, those celebrations of the good old one-to-many technology of print.

"Hey, Mr. E-mail! So! How's the Net?"

"Pretty good."

"So you're basically on-line all the time, right?"

"No, not all the time."

"I don't know where you find the time."

"Believe me, it's easier than you think!" Pause. "You should stop by some time and check it out. I'll give you a tour."

Some of my print friends were obviously not going to have anything to do with my little "tour." But others were feeling the tug of the old anxiety knapsack themselves—the fear of being left behind by progress—and wanted to know more about the on-line world.

"Stop by," I'd say lightly, careful not to sound too much like a Moonie. "You might find you like it."

So friends stopped by, and I took them out for a spin. Despite my efforts to keep my missionary zeal in check, it was hard not to make cyberspace seem as exciting as possible. I couldn't help myself. I felt the pride that anyone feels in showing off his neighborhood. I suppose I also wanted to justify my time on-line. But with off-line eyes watching, I was suddenly aware of how slow things were! What was I usually doing while waiting for these connections to go through? Had I become so used to being on hold for a significant portion of my day that I didn't even notice it anymore? Watching my hands on the keyboard, I was struck by how certain incantations, like my various passwords and frequently used software shortcuts, seemed to have become embedded in the nerves and muscles of my fingers and wrists. Typing them was almost like a reflex action. Also, the notepaper around my desk was filled with mysterious combinations of symbols and letters, which corresponded to Internet addresses and operating instructions and initialization strings, but which some of my visitors seemed to take as further evidence of my having taken a wrong turn somewhere on the highway of life and letters. There were e-mail addresses and gnarly-looking URL's of Web sites scrawled on the outside of envelopes from the few readers who had sent polite, old-fashioned letters about my articles to the *New Yorker*, and whose mail was lying around my desk unanswered.

One of my print friends was a real surfer.

"So does this feel much like riding the old waves?" I asked as we were surfing the Web.

"Not at all," he said emphatically.

I checked my e-mail, moved people around Usenet, took them via Fetch to FTP sites where some of my favorite information was stored ("Look! There's the new version of Telnet!"), went to AOL

("Welcome! You've got mail!"), opened a couple of Telnet windows, logged onto ECHO and the WELL, and, of course, we surfed the Web. But giving an exciting on-line tour was an art that I never quite mastered. Sitting together in front of the small screen made eye contact awkward, and the tension of waiting for things to happen on the screen constricted the conversation. And I could never find anything to show my visitors that was equal to my interest in the on-line world. So much of cyberspace was still invisible, or, like Marley's Ghost, only visible to Scrooge. Eavesdropping on various discussion groups was no way to experience the groupmind; even though I was not yet an avid poster, I knew that. We were trying to use the screen like TV, and compared to TV it was pathetic. Unfortunately, the thing that the on-line world offered, which was the chance to participate in what was on the screen, was impossible to demonstrate in a tour. "I don't know what's taking so long. . . . I guess the circuits are busy today. . . . It's a 'brown-out,' as they say on the Net. . . . Maybe if we tried again after lunch." I could feel my visitors' interest flagging. "You have to imagine what it will be like when there's more bandwidth. . . . " When my print friends were gone I was left with the feeling that in some hard-to-define way they had won.

VII. **The Return of Stephen Gaskin** ◊ ◊ ◊ ◊ ◊ ◊ ◊ ◊ ◊ ◊ ◊ ◊

If there was a turning point in my two years on-line, a Rubicon at whose banks I either had to turn back or keep going forward, it occurred when the groupmind on the WELL became aware of me, and "got into my thing." The occasion for this was an article I wrote about getting flamed that ran in the *New Yorker* in June 1994, in which I revealed some of the strange feelings the flames had stirred in me—indeed, I had pretty much spilled my guts, which was not my usual style in writing pieces.

There happened to be a *New Yorker* topic in the Media Conference on the WELL. To get to it you typed "go media" at the "OK" prompt, then you typed "see" followed by the number of the topic: 748. (There were software settings that provided users with more efficient ways of reading through their favorite conferences, but I never liked to use them, because they made the experience seem too mechanical.) The *New Yorker* topic was one of the most active regular topics in Media. Thousands of words each week were posted about the magazine, written by faithful, knowledgeable, longtime *New Yorker* readers who had strong opinions. The level of conversation was dauntingly articulate.

When I first discovered the *New Yorker* topic I started looking back at comments that were posted around the time the Bill Gates piece was published. As I was hunting through the old responses a nervous-excited but also slightly creepy feeling was jangling the lining of my stomach. I found some nice comments, especially from one pern called <patsy>, whom I immediately imagined as the beautiful hippie chick of my adolescent fantasies, but there were also some negative comments that made me feel like a rubbernecker who slows down and then sees to his horror that he is, in fact, the victim of the accident. I quickly hit "q" for quit, typed "exit," and left the system. Later on, I sent <patsy> a flirty little note thanking her for her post.

After that I sometimes lurked in the *New Yorker* topic to see what people were saying about the magazine, but I never posted in it, and since these were early days yet for me on the WELL, and I had only posted in a few out-of-the-way conferences, almost no one else besides <patsy> knew that <seabrook> was a pern. When "My First Flame" appeared on the newsstand, I nervously checked the topic a couple of times over the next few days, to see if there had been any comments posted about it. I felt weirdly sneaky about doing this, but there was also a strange, secret thrill, the thrill of the lurker, even if you were lurking at your own funeral. There was nothing posted for the first couple of days after the article came out, and then I skipped a day, and then, on the following day, a beautiful summer afternoon in midtown Manhattan, just after lunch on a Thursday, I logged on from the office and saw that a thread had begun.

Topic 748 [media]: The New Yorker
#145: The Sweat of Fear Smells Disgusting (rbr)

Well, good! So here we can turn our attentions to the travails of John Seabrook, who, in a remarkable impression of Ved Mehta, gets ten pages out of one e-mail message.

This piece reminded me a little of Bill Clinton's address to the 1988 Democratic National Convention. Seabrook touches all of the bases, and speaks to many of the points that need to be spoken to, and it's encouraging to see someone speaking to these points under the national spotlight. And yet, he rambles, stumbles over the complexity of the issues, and ends up talking for a very long time for all the actual information he manages to impart.

I knew from my lurking that <rbr> was one of the most energetic and articulate posters on the WELL. With a little poking around the system I had learned he was Robert Rossney IRL, a digital guy who had discovered his writing talent by posting on the WELL, and now did a column about on-line life for the *San Francisco Chronicle*. He had been on the WELL since 1985, and the amount of knowledge and experience about cyberspace he must have accumulated in that time was mind-boggling to me. Clearly he knew much more about this on-line world than I did, even though I was the one writing about it in the *New Yorker,* and this made his criticism of my piece strong.

Then Jon Carroll, the *Chronicle* columnist, posted:

#152: One word: rostrum (jrc)

heh. veteran readers will remember seabrook as the author of the "he likes me! he really likes me!" article about Bill Gates.

This article seemed to consist of the author creating a persona that he believed would be attractive but was in fact intolerable. I found myself rooting for the flamer, altho I'm not a flaming kind of guy. Put this animal out of its misery, I thought, although that was an uncharitable and wrong thought.

Later, when I became friendly with <jrc> on the WELL, I came to understand that his reason for lighting into me like this had less to do with my work than with his own unfocused animus for Bill Gates. He came to regret typing these words before thinking about them (when I asked for permission to reproduce his words here, he responded in e-mail, "I felt as though someone had taken a videotape of me in one of my least lovely moments—blowing up at a desk clerk, say, or storming out of a restaurant—and announced the intention to play it in movie houses across the nation") and I came to forgive him for posting them. But at the time I saw only bloody murder.

Then there was some nervous-excited chatter from other pern— little cries echoing through the jungle summoning the others—then <jrc> posted again:

#160: One word: rostrum (jrc)

I'm sorry; I hate flamewars as much as anyone and more than some. I have no idea why seabrook makes me think evil thoughts. It's just so darn passive aggressive . . .

#161: Kathleen Creighton (casey)

Sounds like he's a chucklehead and will be rightfully driven from the net. In fact, I'd like his e-mail address *right now* myself. I've got this pent-up hostility—oh, 45 years' worth—I'd like to do something with :-).

#162: Avant Garde a Clue (mnemonic)

He's seabrook@well

"Oh, shit!" I said to my screen. This <mnemonic> was Mike Godwin, aforementioned author of Godwin's Law, a tireless advocate and arguer for free speech in cyberspace, both on-line and off, whom I had interviewed recently at the Electronic Frontier Foundation offices, which were then in Washington. I'd forgotten that on that occasion I had also mentioned that I was on the WELL. Now he had outed me.

But no, wait! The posters didn't know I was actually reading this thread. I could go on lurking safely, which was what my instinct was to do. It was a very strange situation, and faced with a strange situation my natural inclination is to lurk. I am a professional lurker, in a sense, and this was the ultimate in lurking: I was actually invisible.

What I did not know, but what I learned in the very next posting in the thread, was that there was a software command on the WELL, the "parti" command, that users could type in order to know when another user was last in the conference. Now another pern helpfully posted this command in the thread, for all to see.

parti seabrook Thu Jun 2 14:35:35 1994 John Seabrook

Then I saw on my screen these words:

So Seabrook. Is the article misrepresented here or what?

I felt as if the television had walked across the room, pulled up a chair, and asked me a question. No, I just want to keep watching, thank you. I got up from my computer and stood in front of the bookshelf for a while and gazed at the spines of the books, but nothing there suggested a solution to this predicament. I sat down at my screen again. I was very vexed—fear, anger, the urge to speak up, and the urge to log off forever were all combined in my stomach. Cold sweat was slicking my palms. Here was Mr. E-mail, who had spent not a little breath singing the virtues of "interactivity"—here, finally, I had the opportunity to sample those virtues, to participate in a discussion about my article with an informed group of my peers. And what did I want to do? I wanted to go back to my summer palace of print. I wanted no part of this discussion, in which my article, which had seemed *sealed* in a sense, by being set in Irvin (the *New Yorker*'s classic font), would here be unsealed, leaving my neat tricks of rhetoric naked, scraggly, and exposed.

I logged on again. <rbr> had contributed another posting to the thread:

#173: The Sweat of Fear Smells Disgusting (rbr)

Seabrook wrote about alt.flame because he went to it thinking that he might find people there who were talking about the phenomenon of flaming. Instead he found people flaming. It was an eminently cuttable graf, among many.

If Seabrook wants to write about his experiences in alt.flame and his paranoia about viruses, and if he does it in an engaging and entertaining way, that's all to the good. I'm not the kind of guy who reads James Thurber's "The Admiral on the Wheel" and thinks oh, for Christ's sakes, put on your glasses already. But Seabrook doesn't write engagingly enough. He's not Thurber, and he's not McPhee, and the subject pretty much demands one or the other. (Can you imagine what this article might have been like if McPhee had written it? Worth close scrutiny, at least.) Like

I said, Seabrook has absorbed, and communicates, a lot of what needs to be communicated; he just communicates it in a painfully uninteresting and self-centered way.

I remembered some good advice from <hlr>'s book: when things get you down on-line, go outside and walk around. So I did. It was a lovely summer day in Manhattan, a day when the air is fresh and the cool breezes from the rivers reach as far into the island as Fifth Avenue. But I was oblivious to the beauty of the afternoon. I was thinking, What is the value of freedom of speech on-line if the result of it is the freedom to be crueler than you would be in real life? Yes, there are constraints placed on our freedoms by civil society—in a thousand ways our natural inclinations are inhibited. But if, in the name of truth, you stripped those inhibitions away and discovered that life was intolerable without them, then where would you be?

On the other hand, the WELL was as free-thinking a society as I was likely to find in cyberspace, and if the condition of that freedom was that I would occasionally have to hear speech that was painful to hear, shouldn't I have the courage to face it, and to respond? But wait a second, this wasn't only about speech. It was also about power, control, and authority. My authority in every sense of that word was being challenged here in this thread. By entering into this conversation, as a pern, <seabrook>, I might be surrendering whatever advantage John Seabrook had as a byline.

On the subway ride home that evening the desire to flame <rbr> was strong. As God is my witness, as God is my witness, I will not take this kind of shit from the pern again. But insofar as I would be speaking as the author of an article about flaming, in which I had taken the high road by not flaming the flamers back, flaming <rbr> might not be a good move, either strategically or spiritually. On the other hand, maybe it was time I just ripped off the suit and wrote. Flame on! No: I would freeze him out. Don't give any of these bastards the satisfaction of a response. I remembered my father's advice in situations like these: Never let 'em see you sweat.

That night, back at home in front of my screen, I logged on again, fortifying myself beforehand by putting Snoop Doggy Dogg on my Discman.

> Now roam in the depths of hell
> Where the rest of your bustass homeboys dwell
> What's my motherfuckin name!

Typing my username and my password, I growled to myself, "Man, You are *bad,* and none of these motherfuckers had better *fuck* with you, because if they do *fuck* with you, you are going to *fuck them up!* Know what I'm sayin? What's *my* motherfuckin name?" I was seeing in my mind's eye the gestures that Snoop makes with his hands in his videos, menacing and contrite at the same time. Stylishly murderous—that was what I wanted to be in this thread. But I was still too timid a dog to post. It seemed like I was more of a cringing golden retriever type than one of the scary-looking Dobermans that loped through a Snoop video.

I saw that Howard Rheingold himself had now shown up in the thread:

#189: Howard Rheingold (hlr)

Let's hope this is the last article in which the writer uses his/her own cluelessness about the Net as a theme. We are at a point in history where cluelessness is not enough; in fact, considering what is at stake, cluelessness is a sin.

Howard! Mr. Virtual Community! Et tu, dude?

#193: Tom Tomorrow (tomorrow)

Howard, I suspect we will see cluelessness as a theme for a long time to come, and I am not sure that this is a bad thing. You have to keep in mind that it is a very small percentage of the population that is as aware of techie stuff as you are.

#194: Howard Rheingold (hlr)

I made a distinction between naïveté and cluelessness for a reason. . . . When you are a writer on assignment to the *New Yorker*, it might be assumed that you, like many others, are ignorant about a topic when you begin to explore it; to propagate idiocy regarding e-mail as viruses, particularly after making a big deal about your interview of one of the world's experts on the subject, is, IMO, aggressively and dangerously stupid, not charmingly naive.

To flame, or not to flame: that is the question:
Whether 'tis nobler in the mind to suffer
The slings and arrows of outrageous assholes
Or to take arms against a sea of troubles
And by flaming end them? To rant; to rave;
Evermore; and, with that flame to say we end
The heart-ache and the thousand natural shocks
That the pern is heir to, 'tis a consummation
Devoutly to be wish'd.

At 2 A.M., I checked the topic one more time before going to bed, and saw:

#202: Corrigan (patsy)

Seabrook? What do YOU think?

The next morning I got out of bed with the resolve to make one posting, then forever deny these buttheads the pleasure and benefit of my company. I would pack up my high-tech geodesic dome and find another system to roam. But before I left, I would respond.

But how should <seabrook> speak? I opened a Microsoft Word window on my PowerBook and stared at the blank space. I typed a few sentences inside the window. Intimate shorthand was fine for e-mail, but now I was speaking in public, addressing my critics. I wrote a few more sentences, but they came out sounding like Mark Antony addressing the Roman Senate. I dragged my cursor across the sentences and hit "Delete." I clicked on another window and went to work on the article I was writing, but little thoughts for posts kept fizzing up like Alka-Seltzer tablets in my mind. Finally, I logged onto the WELL and typed:

#206: John Seabrook (seabrook)

I am reluctant to post here only because whatever I say will be an attempt to spin the discussion in a way I want it to go and therefore make it less interesting (at least for me). Please don't mistake my choosing not to post here as a sign of lack of respect for you all.

But this attempt to retire into the cool shade of print was unsuccessful.

#214: blue galaxies (robbg)

John's response in 206 is a little disappointing. One of the gifts that the WELL and other systems offer writers is the chance for engagement; John's using it to be an observer. It is, of course, his right—and he may see it as his duty as a writer. But I can't help but wonder, since he's opted for such a highly personal route in his articles, whether he's really serving his readers.

Jon Katz, a byline I knew from print, who was one of print's few apostles for on-line, now showed up in the thread:

I think we're getting another delicious look at how one culture—journalism—is encountering another, almost bumping into it. I think it will be eons before writers and reporters for our best media can just unselfconsciously listen to, talk with and kick ideas around with readers. I thought John Seabrook's article was an interesting, fun account of his experience, but I think <seabrook>'s posting is just coy. It suggests we will all be influenced by his perspective. It also suggests the only real posting is an article or an op-ed piece. Postings here don't really count. I was sorry to see that. Just do it, John.

Okaayyyy, I'll try again. It still felt strange, writing a post. Rhetoric—dualism, hyperbole, various other tropes—that I would have employed naturally in writing for print sounded false, somehow, in this context. A different, more spoken language was called for. The readers were closer. Print is like a stone with a message tied to it that you throw across a chasm to the readers on the other side; on-line was more like a bridge across the chasm, or at least a rope. So print writing tended toward being a monologue, on-line writing toward being a dialogue.

It also felt strange because writing on-line was the way you engaged other people, whereas IRL writing was for me a way of withdrawing from the world, something you did by yourself.

#228: John Seabrook (seabrook)

I'm not saying (as #214 seems to suggest) that I am not interested in participating in any discussions here and thus not getting "the essence of life on-line." I am looking forward to finding out what the essence is. But citizens, let's remember that this particular conversation is about my writing, which is sort of like it's about me. It's kind of strange. I've never heard a conversation like this before. Put yourself in my position. It's kind of scary. It wasn't that long ago that I was tremblingly showing my overwritten manuscripts to editors and feeling that sick feeling in the pit of my stomach when their voice came onto the phone with that sympathetic tone in it saying, "It's really well written but . . ." Now, I'm listening to people call my writing/me "intolerable" (jrc); a "chucklehead" (casey); "painfully uninteresting" (rbr); "a specialty writer" (kj); "not a specialty writer" (mnemonic); and "aggressively and dangerously stupid" (hlr) . . . But perhaps the cruelest cut of all was to compare me to Ved Mehta (rbr).

#229: Howard Rheingold (hlr)

It's an initiation ritual, John Seabrook. Stick around and help us dump on the next guy. ;-)

Another poster said:

Welcome to (if I may say it) the real world.

I responded with another posting, then read it closely, worried that my written words would inadvertently reveal more of my hostility toward some of these posters than I wanted to reveal. In speech, it's relatively easy to say one thing and be thinking another, but written words seem to have a more symbolic relationship to your thoughts and emotions, especially when the words are written in haste. You are indeed a kind of "mental nudist," as <tex> had remarked somewhere on the system, when you started writing on-line.

#243: John Seabrook (seabrook)

RE "welcome to the real world." Thank you, but actually I've been there before. I have heard plenty of criticism of my work and worldview. This is not like that. This is more like having the book group over to your place and hiding in the couch all evening.

It's a little like that old fantasy of being able to go to your own funeral. I'm the invisible man. I've mixed up this wacky science experiment in my basement and taken a sip of the smoky potion and doggone it I'm invisible now. They can't see me! This is not the real world.

But wait, how do I get back to the real world?

#246: David Gans (tnf)

John, it is your decision and yours alone to hide in the couch.

#285: One word: rostrum (jrc)

JADP [Just a Data Point]: Stewart Brand, over the news conference, reports that the fact checkers saved Seabrook's ass in the small section in which he quoted Stewart which was, in its unchecked version, according to the Steward, "hash."

This post unnerved me, and I immediately sent victim mail to <sbb> (which I fear may have begun "Stewart, O my Brother"), and then posted in the thread:

#293: John Seabrook (seabrook)

I don't understand the purpose for posting #285 by jrc. Is this an attempt to discredit me? If it is, I don't understand what jrc has against me. Because he's the host here his remarks carry a certain weight. Does he not want me around here? As I've said before I'm way too manly to let this stuff bother me personally, but I'd like to try and understand the situation.

Strange memories of boarding school were going through my head. I was back inside that big open room again. I remembered clearly what it was like to lie there in my alcove listening to the mean silence, the tension so concentrated that it was like a tick crawling on your skin. Boys would swoop down on you like barn owls and, say, hold a tennis racket to your stomach and, with a stiff brush, stencil a bright white net into your stinging scarlet flesh, then rub toothpaste over it.

I had supposed in my naïveté that the groupmind was going to be some nice Emersonian idea of an Over-Soul, not realizing that the groupmind could also be Stephen Gaskin, right in your face, screaming at you as Gaskin had screamed at John Coate on The Farm:

> "You're Jack the Ripper. You're a SON OF A BITCH!!! You have the morals of a snake. Now get out of here!"

The question was, which is better? A charismatic leader like Stephen Gaskin or the groupmind? A single human mind, or Mind? In theory, the groupmind was a more perfect leader than Stephen Gaskin, because it had a self-correcting mechanism built into it. Although it could screw up from time to time, it was harder for the groupmind to go completely around the bend. The "system" would not let that happen. On the other hand, the groupmind was poorly organized, easily distracted, hysterical, inaccurate, and insecure, especially about slights from the one-to-many world, although at the same time it was tenderhearted and quick to forgive such slights. It was God, but it was also a mob, and it could behave like a mob sometimes.

I got up from the screen and walked down the street to the deli to get some beer for the upcoming Knicks-Pacers play-off game. (Live sports was pretty much the only thing I used TV for anymore; on-line had replaced all of TV's other functions for me.) When I came back I posted:

#312 John Seabrook (seabrook)

I just walked down Greenwich Street to get some beer (Go Knicks) and on the way back thought of another member of my on-line dream team—jrc as waterboy.

#351: Howard Rheingold (hlr)

Seabrook, the painful and necessary question to ask yourself when people dump on something you have written is: "Is there useful information in here? Could I learn from it?" Granted, it isn't easy to calm down and not get all twitchy and defensive when strangers attack your writing. But I've seen a lot of stupid know-nothing nasty criticism on the Net, and I've seen some intelligent, well-informed nasty criticism, and I have found it worth the pain to learn from the latter (and just let the former slide—there's no sense getting in a pissing contest with a thousand idiots who have time on their hands).

The thrash lasted about ten days altogether. I had thought the anger would go out of it as soon as I began posting in the thread, but I was wrong about that; instead it escalated and became a full-blown thrash. One pern wrote a parody of my lead; other pern reprimanded me for my snide responses to certain postings. The anger only began to ebb when <sbb> himself made an appearance.

#360: Stewart Brand (sbb)

Let me quickly agree with Mike Godwin and others that there is no flaming in this topic. But there is something else that doesn't have a name yet. When I was on the receiving end of something like it (which also was not flaming, technically), the terms that came to mind were Gang Bang (of the unwilling kind) and

Chicken Peck—where one of the flock shows a bit of blood, and a few of the other chickens (it doesn't take many) use it as a target to peck the bleeder to death. "Piling on" is a term that has been used correctly a few posts back.

The earnestness and *energy* of the critics—eager and public-spirited about pointing out the errors of their comrade, for the good of all—is breathtaking. It gets exquisite when the victim is mocked for "being so defensive."

Brand's posting caught one thing that was strange about the situation. I was incredibly pissed off at some of the posters, but they, the other pern, believed they were *helping* me. The key seemed to be somehow to put aside my wounded feelings and try to join them in getting at the truth, whatever that was. I thought of another posting of <fig>'s I had seen in Archives, talking about life on The Farm:

We were really into the enlightenment thing, at least in the first 6 years or so. I guess we got sort of fatigued at striving for it after that. Especially popular and emulated were tales such as the one where the monk, seeking to meet with a famous master, repeatedly asks for permission to enter the gates of the monastery only to be rudely turned away by the gatekeeper. Finally, in frustration, he tries to enter the gate but the gatekeeper slams it shut on him, breaking his leg. He is instantly enlightened!

But I was not that enlightened yet.

#462: John Seabrook (seabrook)

. . . I've started having these hallucinations about what Herman Melville might have heard if he'd gone on-line after publishing *Moby Dick*. . . .
[Excerpts from the Melville thread.]

. . . Harpoon THIS, asshole . . .

. . . you have done a serious disservice to whaleboat captains everywhere by portraying one of our number as a weirdo. FWIW, most whaleboat captains I know are decent, caring individuals who . . .

. . . I can't believe your editors didn't correct this huge, glaring error in your book: you made the whale, which is supposed to be evil, *white*. . . .

One evening toward the end of the thrash Lisa and I were riding through Soho in a cab. I thought of a devastating reply to an incredibly annoying posting I had read before we left the house that evening, and I tried it out loud, under my breath. Lisa said, "Are you talking to those people on the WELL?"

"Actually, yes."

She gave a small, somewhat sardonic-sounding snort and looked out the window. I said, "I'm sorry." We rode in silence for a while. I said, "In the fall the Internet was always there, but we did not go to it anymore."

"Who said that?"

"Ernest Hemingway."

"I didn't know Hemingway was on-line."

"*The Sun Also Rises* was written about the Internet. The hero gets his wound on the Internet."

The thrash ended decently. There was a friendly post from a pern named <booter>.

#407: Festive Circumstantial (booter)

I have watched seabrook go from a freshly irritated novice to a hardened cluemeister in one topic.

We should rename this topic "The Defloration of Seabrook."

He may be a newbie, but he has a clue firmly clenched in his wet little fist.

Nice work, dude.

That cheered me up a lot. What a nice pern!

And then, in the summing-up phase of the thrash, there was this post from <mariana>:

#428: Mariana Rexroth (mariana)

Following this thread has been most interesting: first, no Mr. Seabrook . . . you wouldn't even know he was around; then, Mr. Seabrook as the Outsider (somehow Colin Wilson comes to mind, but that's just my brain playing funny tricks); and then . . . hooray, <seabrook> as one of the gang. WELLcome to the treehouse. I suspect you just might find it a good place to hang out. I mean it: I'm glad to see the shift . . . I was worried there for a while, but the final result confirms my sense that this medium works . . . maybe slowly, as any self-correcting system works . . . and works in the best way, a way that allows people to reach their own level of comfort and understanding.

That was true, too. The system *had* worked. I was no longer an outsider. I was one of them now.

Chapter Six   The Post and the Thread

I. **Mad King Bruce** ◊

Now the newspaper pages of all my WELL days fly off my screen.
I get a "pseud." I begin posting in different conferences. I do what
I can to "build community." When I go to the Hole show at the
Academy I write a quick review of it for the Music Conference.
Then I write a second review, of the Beck show at Irving Plaza. I
discover the pleasure of immediate "publication" and feedback. I
am not getting paid, of course, which I might be if I were writing
this for print, but I don't have to work as hard. And no editors!

I play little adult games with other people. We'll take a public
position in the threads—i.e., both defend another poster in a thrash
—and then send each other private e-mail about how lame the poster
is. I discover the value of spreading compliments around—a "Right
on!" here, a "Great post" there—so that the next time the
groupmind lights into me, I'll hopefully have allies. Politics, diplo-
macy, all the "polite" arts of society begin to grow like kudzu. On

the WELL, as everywhere on the Net, everything grew like kudzu. You got entangled, then more entangled, in a fine spun mesh of platonic or not so platonic relationships.

The WELL was full with liberals, which was okay with me, although at times one encountered a certain New Age kind of liberalism that I don't think Jonathan Edwards would have cared for very much. When the pern started sending each other "beams," which were WELL units of positive energy, or calling for "Group Hugs," I usually logged out. However there was much more to the WELL than beams and group hugs. The WELL was the closest thing to a *functional* utopia of free speech of any place I encountered in my two years before the screen, and as such it was a very interesting place. On the WELL, as everywhere in cyberspace, one lived entirely by acts of speech. You fought, made amends for fighting, loved, gave nuggies to your little buddy, all by acts of speech. You ate by speech, in the sense that you absorbed information from one place on the WELL and turned it into speech in another place. The intimacy of the ASCII text, combined with the lack of social boundaries on-line and the relative safety of the screen, which allowed you to emote without danger of a violent physical response from your audience, tended to make one's relations with other pern get personal pretty quickly. In one of his more incisive postings about whatever in the heck it was we were doing together here, <rbr> had written, "What is being transacted here is a series of personal relationships." (One way of using the THING was to refine one's vision of what the THING was.) When I read that remark, I wanted to post something sincere like "Brilliant Bob!"—but because I happened to be pissed at <rbr> for some other postings of his, I didn't.

But what was it we were doing here? That was what kept me logging on four, five, six times a day. In one sense, the WELL was a beautiful idea. Here, at the geographical end of the American frontier, a new frontier idealism had sprung up. As <fig> had written:

> It is my assertion that the actual exercise of free speech and assembly in on-line interaction is among the most significant and important uses of electronic networking; and that the value of this practice to the nation and to the world may prove critical at this stage in human history. I regard the WELL as a sample of the kind of small, diverse, grassroots service provider that can and should exist in profusion, mutually accessible through the open channels on the Internet.
>
> The possibility that the future "Internet" (or whatever replaces it) may be dominated by monolithic corporate-controlled electronic consumer shopping malls and amusement parks is antithetical to the existence and activity of free individuals in the electronic communications world, each one able to interact freely with other individuals and groups there.

But although the WELL was an excellent place to talk about democracy, it wasn't actually a democracy itself. In what John Perry Barlow liked to call "meatspace"—the hopelessly postindustrial but stubbornly persistent Real World—the WELL was a business. In a constitutional sense, a pern's relationship to the owner of the WELL was more like a New York Yankee's relationship to George Steinbrenner than a citizen's relationship to the head of a democratic state. Earlier in the life of the WELL, when it had been co-owned by the Point Foundation and NETI, the divide between the community's values and the management's values was not very wide; in those days the community *was* the management, more or less. But with the beginning of the Bruce Katz era, which was just getting under way around the time I joined the WELL, and which happened to coincide with the explosion of business interest in the on-line world generally, the gulf between the owner's interests and the community's interests became more pronounced.

Katz had inherited the Rockport Shoe Company from his father, built it into a thriving enterprise, and sold it to Reebok for $100 million in 1992. In buying the WELL for a total of $4 million he was getting one of the oldest, best-known computer conferencing systems in the world, and it had more recently become an ISP, at a time when Internet euphoria was starting to make some people a lot of money. Naturally Katz was looking to increase his investment by "growing" the WELL. He hoped to market the WELL as an up-scale, literate, community-oriented alternative to the Big Three, a system that would offer an Internet connection, but also intelligent conversation and interesting, useful people. His plan was to start a number of "regional WELLs," or "McWELLs," around the country, each of which would hopefully develop their own local community feeling, but which would all be linked together into one big "global" WELL. He was adding new lines and equipment to reduce busy signals, improving customer service, and offering PPP software kits, and he planned to lower prices.

But when the groupmind began to hear reports of Katz's plans to grow the WELL, it had an anxiety attack. The Owner was as enlightened a businessman as one could reasonably hope for, but was his relationship with the users different in any essential way from the relationship he formerly enjoyed with his shoes? What many in the community dreaded was that Katz would turn the WELL into another AOL, with brainless posters lobbing witty remarks like "Rap Sucks!" or "Metal Sucks!" back and forth across the wires all day long. When it was reported in the News Conference that Katz had commissioned a team of software designers and spent a million dollars to design a GUI to replace, or at least supplement, the old, unsexy but lovable keyboard command-based interface of Picospan, and that, moreover, Katz wanted to make the WELL into a major

Web publishing site, it seemed as though the darkest fears of the groupmind were going to be realized. Somewhere between the ten thousand users of the present and the million users of Katz's fantasies, this quaint white-washed little village of words would turn into the Thai island of Phuket—i.e., overdeveloped. At the very least you would end up spending most of your time reading postings from people you didn't know and didn't want to know. And, in spite of the lip service we occasionally paid to the many-to-many ideal on this little commune of ours, hardly anybody really wanted to let the unwashed masses in.

Some of the oldpern like <tnf> and <hlr> had talked to the owner f2f about the community's concerns, but the meetings did not go well, and these and other pern expressed their frustrations with Katz publicly on the WELL. Katz *was* difficult to talk to. He seemed less interested in listening to your ideas than in telling you his. During my only f2f with the Owner, which occurred in the company of several other pern at Trattoria del Arte, a restaurant across from Carnegie Hall, in New York, in March 1995, Katz spent an hour talking almost nonstop about plans to grow the WELL, especially the New York WELL, tossing out wild, half-baked, almost desperate-sounding ideas, then abruptly segueing into sad musings over all the money he was spending on his new GUI. One of his ideas was to create a troupe of "Cyberscouts"—kids he envisioned doing good deeds on the McWELLs around the country, like creating Web pages for little old ladies, and maybe going from door to door selling subscriptions to their regional WELLs (whether they would wear uniforms was still up in the air), just as the boys and girls of my youth sold seeds and cookies and greeting cards (would cyberscouts swear on Scouts' honor not to flame?). It was nice that the Owner's marketing urge was mixed with a sense of social responsibility, but rather than being enlightened by his altruistic impulses, he seemed confused by them.

Then suddenly, because he was late for a Tibet House benefit across the street at Carnegie Hall, Katz gave the waiter his credit card, and absentmindedly rushed out before the waiter returned with the unsigned bill.

One of the people at the table said to the waiter, "Well, Mr. Katz left, but we'll give you a really big tip if you'll let us order dessert and after-dinner drinks and coffee and sign the bill in his name." But the waiter wouldn't go for it.

It would have been a small victory for we the pern.

Perhaps the greatest of all Bruce's sins, in the eyes of the users, was that he didn't really participate in the community of the WELL. It could have been because he was too busy running the WELL, but the groupmind was very sensitive to the perception of slights of this kind. Because <katz>, the pern, had not bro'ed down with the other pern, Bruce, the person, was suspected by some of being an elitist, or at least of just not "getting it," and therefore of being unlikely to act in the best interests of the community. His occasional appearances on-line had the aspect of a monarch waving woodenly to his subjects from the balcony of his palace. He reminded me a little of myself, before I posted in the *New Yorker* topic.

Clearly, it was only a matter of time before the groupmind got into <katz>'s thing. This happened in a big way in September 1994, in a memorable thrash in News that began when the Owner announced he was relieving <mo> of his responsibilites as General Manager of the WELL and was taking over the job himself. As it happened, the WELL was then experiencing a series of particularly frustrating system problems. The pern were troubled, and the Owner tried to answer their concerns. But he sounded defensive—and that was the target the others pecked at.

#1057: Bruce Katz (katz)

Why do I always feel so lousy reading this kind of stuff from you people? I had been reading this religiously but it kinda disintegrated around posting 800 or so. I lost interest and felt like the bullshit was just flying around again. Sorry if I seemed peeved but it gets very tiring to have everyone talking as though a businessman is by definition a louse. I am one of the founding members of Businesses for Social Responsibility and I go to their meetings twice a year. It is a child of the Social Venture Network of which I am also one of the founding patrons. We are a group of individuals that believe that the proper role of business is to build a just, humane and sustainable world. Working at the WELL is my way of carrying out this mission. I believe in this medium and what it can do to help empower social change from the grassroots level. At last a truly democratic medium. In the future as we move into other parts of the country we will work to make this avl. to as many people as possible from all walks of life.

Last time I looked it was me writing the checks and taking this bet. If you don't like the new stuff don't use it. If you don't want to play in a bigger place keep your configurations set to the WELL you're in and you'll never notice the other people populating the Paris WELL. It's up to you. As to all these accusations of not working on the problems I can assure you that I am. This is a very tough job at the moment. Please understand that I do get a lot of what this community is about. I have learned a lot from this topic and have already incorporated some ideas that have come up. I am sure that I will reread it again in the coming months to glean more. I wish I knew how to keep the tone more positive and constructive instead of all this nay saying and distrust. I don't believe that I have done anything since I have been associated with the WELL to harm the community. It is extremely hard to put all of your business strategies on line although I understand the apprehension. I only would ask that you give me the benefit of the doubt a little more. I hold myself to some pretty high standards and it is hard for me emotionally to work as hard as I am and to take the kind of denigrating flack that I do in this topic. Have a little faith. It's all that keeps me going some of these days.

#1066: You can always count on (jstraw)

So Katz is off with his thumb up his ass while all hell breaks down on a 10,000 user system. He wants a million. Laughing my fucking ass off, I am.

#1128: private-conf organized adulation (rbr)

What I would like Mr. Katz to understand is this: there are many people here who use the WELL to practice community.

Many of these people have been doing so for five or ten years; it is often an extremely important part of their lives.

To some of us it is as if someone who lives on a hill and who comes down from time to time to see what is going on in the churches that he owns has decided, from consulting with his advisers and feeling the pulse of the nation, that church-going is going to be an important thing across the land, and that he would really like to make this happen, and make a little money, by building more churches, and infusing these new churches with some new ideas about faith and service and community that he's picked up along the line.

The intentions are obviously good. Unfortunately, this person doesn't go to church himself—well, he has his own church, maybe, up on the hill, but down in the valley where people go to church three or four times a week, he hasn't been seen at the services. True, one of his advisers occasionally pulls up in his Maserati, sticks his head in to see what all the yokels are talking about, and drives back up the hill, but all this does is make the parishioners more nervous about the future of their church than they already are.

There is a community of souls here who have an emotional stake in the future of this place. These people are open-hearted, for the most part, but they are also suspicious, and some are jealous. They give their respect easily, but not freely: it must be earned through good words and good deeds.

So far, the words have been unconvincing to many, and the deeds have been downright disheartening—you fire the technical manager, proclaim yourself the new manager, and skip town, and while you're out of town the system goes to shit because of something that could readily have been predicted and prevented. If the WELL had any competition in the weird little market that it serves, they'd be pissing themselves with glee over a stunt like that.

For everyone's sake—for your own, and for ours—please: try to develop some understanding of and sympathy for the importance that the WELL occupies in the lives of many of its users. If you could bottle something like that and sell it, you'd make millions. But you're not going to be able to do that if you don't understand it, and it is abundantly clear that your understanding has a long way to go.

How would our elitist leader respond to the reverse elitism of this post?

#1141: Bruce Katz (katz)

I would like to respond to this notion that I fired the technical manager and went away. I hired Maurice. And I had to ask him to step aside as the operating manager for legitimate reasons that in deference to the man I refuse to discuss here. This was not some action I took lightly. I like Maurice and it was tough as hell for me to have to make the decision I did.

Keep in mind that it has been me in the background constantly pushing to make sure we have the latest modem equipment and that there is no busy signal etc. Maybe we should reboot the sequent and give you back the 2400 baud modems and see if that's better.

Oh yes and in reference to some of jstraw's posts recently I would like to say with all due respect, Fuck you too!

I thought to myself, Right on! Dude! Welcome to the treehouse! But Katz was embarrassed by this posting, and later, when I e-mailed him to ask for permission to use it, he e-mailed me back, "After over 1110 postings in this topic, trying to explain my still evolving plans to the community at large, I was clearly beginning to lose patience at this point. You can only freely commit yourself to a stoning for so long before even the thickest skin begins to yield. I feel embarrassed to have written this, and give permission for it to be reprinted only in the hope that others might learn from what took place. Am I just another member of the community as some in this conversation would wish, or am I a proprietor of a service company providing the platform and tools for this dialogue? I find this to be the most difficult task in trying to run the day to day activities of the WELL operating company."

#1146: Clothes by Foreign Host (jrc)

Isn't just a bit tacky when the boss starts saying "fuck you" to the customers? Particularly when the boss is a multimillionaire garment and shoe mogul and the customer is a 33-year-old photographer from Topeka. Doesn't, I dunno, noblesse oblige figure into this somewhere? Or maybe courtesy, generosity and common sense?

#1170: John Perry Barlow (barlow)

There is simply no pleasing you people. One minute Bruce is the yoke of slick-shoed Capital come here to mall-ify our village. Then he says something any man might say after weeks of being attacked by ducks and immediately everyone's shocked and calling for decorum as befits a proper CEO.

You guys are just in for him. He can't win. I hope you will be satisfied when he decides that the emotional overhead is just too high here and sells the whole thing to somebody who really doesn't give a flying fuck.

And anybody who wants to call me a sycophant should have the pleasure of doing so to my face at the WOP [WELL Office Party] tomorrow evening. I will hand you your ass, sir.

#1189: The WOP Dominatrix (kls)

Barlow, in addition to being a sycophant, you're a thug and a cretin. Your threat of physical violence is entirely what I'd expect from a person with your limited intellectual and emotional resources. You may be assured that if you lay a finger on me, it and the rest of your sorry ass will land in jail.

#1211: cognitive dissident (jstraw)

I honestly don't mind your telling me "fuck you." It was the first truly personal act of communication I read from you, to anyone.

The fearful self-righteousness that our Owner had to tolerate! The absolutely infuriating hypocrisy of liberals like us, who talked such a good game about many-to-many, but who, given the opportunity to let the many into our little world, turned out to be mainly interested in protecting our turf! It was almost too painful to watch—though all the more compelling as entertainment because of that.

II. **The Groupmind, Pt. II** ◊ ◊ ◊ ◊ ◊ ◊ ◊ ◊ ◊ ◊ ◊ ◊ ◊ ◊ ◊ ◊ ◊

As was true of the Net at large, the WELL was an anarchy. Apart from the requirement that your identity be available to other people, the only rule that was written down on the WELL was "You Own Your Own Words," which appeared on your screen every time you logged on (and what that actually meant was the subject of a spectacular thrash that was stored in Archives). The only authority figures in the community were the "hosts" of the conferences, but their function was more custodial than authoritarian. In matters of real importance to the community, the many-to-many always decided. There were no rules about what you could or could not speak-act; there were no civic ordinances, no cops, and none of the thousands of other checks that exist to prevent people from running amok in the real world. If you ever asked anyone the rule about something, you were liable to receive the answer the only rule is that there aren't any rules. And it all worked amazingly well.

But it was a newbie mistake to suppose that just because the rules were not written down, they didn't exist. Ah, you naïve pern, you. Why do you think that this particular anarchy *was* so peaceful, that it was almost free of the kind of random sociopathic behavior by individual posters that groups on AOL and Usenet were prey to? It was because there *were* rules. They were the unwritten rules of many-to-many culture. The poster did not have to obey them, of course, but if you did not obey them, you would not get as much back from the screen.

The community of the WELL was in this way more congregation-alist than democratic, more like a church than a state. It was a congregation of believers in virtual life, and although the groupmind was canny about "copping" to the mystical aspect of itself, you saw it in certain threads, especially when a pern whom you knew only on the WELL died IRL, and you found yourself crying real tears at your screen. (But just as death could bridge the distance between people, it also could expose the limits of the medium. A user was dead, grieved by friends and family—blank postings were a popular show of grief on the WELL—but to you it was merely as though a blip had disappeared from your screen.) It was a church of free speech and equality, yes, where the litany, "freedom of speech, even speech that you hate, freedom of speech, even speech that you hate," was chanted at vespers, matins, and nones. But the rights of the individual posters were not protected by laws; instead they were defined by the dynamics of the group.

The fundamental dynamic, which was present in virtually all public discourse on the WELL, was that of the post and the thread. In the individual posting, honesty and directness were encouraged, and personal style, too, within certain limits. So long as you threw in an occasional IMHO, you could pretty much say whatever you wanted to say (though certain popular posters, known on the system as "pets," could get away with more than others). But complete freedom of speech was granted to the individual poster only inas-much as the poster was accountable to the thread—to the totality of all the posts, and the ineffable flow in which they occurred. And the ethos of the thread was not the same as the ethos of the post. It was precisely because you *could* say whatever you liked in a posting that the authority of the thread in which you were speaking-acting had to be particularly strong. The thread was fragile, emperiled by the wilderness of antisocial behavior all around it, just as the seventeenth-century colonies in New England were emperiled by the wilderness all around them. Just beyond this pleasant society of mind, where there was wisdom, sympathy, and useful information, there was a howling void of savage shitheads willing to take full advantage of the anonymity and freedom from ordinary responsibilities that the on-line world offered. The void was always out there and the thread was always aware of it.

Therefore, although in theory you could say anything you wanted to say on the WELL, in practice you often found yourself posting things like "So and so's post is right on," or "What he said," or "This is the best topic on the WELL!" The pressure to conform to the thread was strong. It was fed by the general desire to belong, to find a home on-line, as well as the desire to seem "clued" to the others in the group. The WELL was a pretty homogeneous place to begin with, demographically speaking, and the web of oneness was always being tightened by the constantly spreading memes—language, custom, jokes, software expertise—that bound minds together in a thousand ways, producing an ever-stronger groupmind that the individual poster opposed at his peril. In this respect, although the WELL was rooted in the idea of social equality, in practice it tended to promote something like the opposite of that ideal—it divided people into insiders and outsiders. This phenomenon was by no means unique to the WELL, but could be observed in many other sites across the Net where virtual communities formed.

On the WELL the individual poster made a kind of unspoken covenant with the groupmind. You the poster agreed to do your bit to help feed Mind with your speech, to speak honestly, to add knowledge to the system. In return for that you got a peaceful society of Mind on your screen. You were able to lay back in the cut, as the rappers say, to suspend yourself in the flow of the thread, and harvest ideas, companionship, and "recognition" from it, and that felt good. But in return for that state of grace you agreed to give up a certain amount of the pure individual freedom that was in theory the promise of cyberspace, and was practiced for better and worse out on Usenet. You would submit to the groupmind's authority, even though the groupmind might do its will according to a system of "justice" that you didn't understand, that might even seem cruel or arbitrary to you. You came to accept that your own personal conflicts with the groupmind were essentially futile, symptoms of your own grandiosity which would disappear when you had been around the system a little longer, and had achieved a higher level of enlightenment.

Although the groupmind was rooted in the concept of the many-to-many, it was actually a conceptual leap beyond it. The groupmind anticipated the central flaw in the many-to-many concept, which was that the many were generally not worth the time one invested in listening to them, and sought to solve it by creating an atmosphere in which the motive for stupid, time-wasting, and antisocial speech was eliminated, or at least so strongly discouraged that it was virtually eliminated. The question, of course, was whether the individual poster was indeed freer within the context of this medium than he was as a passive consumer within a one-to-many medium, or whether he had merely traded one form of control for another.

From time to time the frustration of having one's idiosyncratic impulses planed off by the groupmind would inspire the poster to burst into what was known on the WELL as a "rant." The groupmind had an amused tolerance for rants (depending on who was doing the ranting). It recognized rants as a necessary and useful way of exhaling gas that might otherwise prove noxious to the commonweal. And the fact that anyone could tear off on a rant at any time was a proof that this was a liberal, free-speaking society after all. But let the individual challenge the collective consciousness in a serious, substantive way—let the poster wonder aloud the wisdom of allowing a crowd to be God—and the groupmind would react with a speed and a vengeance that would make that pern's head spin.

At the same time, in order for the system to self-correct, and not slide off into a benign form of totalitarianism, some posters *had* to challenge the authority of the groupmind once in a while. This was what led to the thrashes that in turn supplied the energy that drove the system to produce more Mind. (IMHO!!!) But few posters had the stomach to relish a full-blown confrontation with the angry groupmind. The ritualistic group scapegoatings and ghastly virtual bloodlettings that were stored around the system were a bit like the heads of traitors put up on the Tower Gates in medieval London, to remind citizens to obey. And even if a poster did have the nerve to take on the groupmind, the enterprise required a lot of time and energy. The brave poster would log on first thing in the morning to find himself flayed, pilloried, roundly abused, leaving him momentarily *farklempt:* a bad way to start your day. You had to be on-line for a while before you got used to the force of harsh words, flung at you on the screen. Then, after a while longer, you came to look forward to it, and to miss it when it was gone.

III. **User** <z> ◊

When it was focused on solving a specific problem, the groupmind was a beautiful thing to behold. In the Parenting Conference on the WELL, for example, the groupmind was consistently at its best, answering questions from worried parents about weird rashes and croups, offering solace, making the knowledge flow. The groupmind's breadth of knowledge consistently amazed me. On several occasions, having posted in the Books Conference a request for obscure information about an author, or in the Grateful Dead conference about a version of a song, I received a correct, informed answer back within two hours, and I could only post,

O Groupmind you are indeed wise.

The groupmind could also be kind, caring, and compassionate. Rheingold's *The Virtual Community* contained a story that had become a kind of parable ("WELLtale") on the system, about a man whose daughter was desperately ill, being kept alive by machines; her father was sitting next to her bed with his computer, pouring his heart out to the folks on the WELL.

Woods Hole. Midnight, I am sitting in the dark of my daughter's room. Her monitor lights blink at me. The lights used to blink too brightly so I covered them with bits of bandage adhesive and now they flash faintly underneath, a persistent red and green, Lillie's heart and lung.

And the WELL folks were there for him. Rheingold wrote of "sitting in front of our computer with our hearts racing and tears in our eyes, in Tokyo and Sacramento and Austin, [as] we read about Lillie's croup, her tracheotomy. . . ." The little girl recovered and later the man wrote about his experience in the *Whole Earth Review:* "Before this time, my computer screen had never been a place to go for solace. Far from it. But there is was. . . . Typing out my journal entries into the computer and over the phone lines, I found fellowship and comfort in this unlikely medium."

But it was a mistake to think the groupmind would always be sympathetic. Take the case of user <z>, who joined the WELL in 1994. She soon became a co-host of the Peace Conference, and started a thread in the private conference for hosts, Backstage, which only hosts were magically empowered to enter, to solicit ideas about how to revive Peace, which had been sort of a bummer since the Gulf War ended. Posters offered good advice; <z> responded with fulsome thanks to each suggestion. The revelation that <z> had become a host so soon after joining the WELL led to mild flamey outbursts in the thread by a pern named <diogenes> (that was his lantern), decrying favoritism in host selection, which was a tributary of another raging thrash that appeared from time to time on the system. But for the most part the groupmind was on its best behavior during this first <z> thread.

Then, in late August or early September, <z> had a battle with her co-host in Peace, and so was no longer a host, and thus no longer allowed in Backstage. This led to a remarkable posting that began a second thread, in the Hosts conference, lamenting her misfortunes in Peace. In this post she repeatedly referred to herself as being on "the doorstep of the Back," that is, hovering at the threshold of the Backstage Conference, but no longer being able to go there, where it was so warm and the people were so friendly, and how sad she felt about that. It was a powerful, sort of crazy posting. It tapped into something raw and true about the on-line medium, which was the bliss of having the flow for a while but the sadness of losing it, and then experiencing a form of exile.

In both threads <z> repeatedly posted about how wonderful she thought the WELL was. She got on-line, read the story about the man with the sick daughter in *The Virtual Community*, and fell for the whole idea of the congregation hook, line, and sinker; she was gut-hooked. If a fish gets hooked in the side of the mouth it can hope to tear the hook out or spit it, but once it swallows the hook it's hooked in the gut and then there's no hope for it. Indeed, some of the other pern in the thread, who were perfectly capable of gushing about virtual community themselves in other circumstances, and even calling for Group Hugs, seemed to be a little taken aback by the vehemence of <z>'s enthusiasm for the wonderful new community of the WELL.

A month after that second thread ended, <z> started a third thread in another conference to talk about how depressed her husband was, and to ask for advice. The groupmind learned that <z> was around fifty or so and a writer. Her husband was a good deal younger, a computer programmer whom <z> seemed to sort of worship for his skills, but he was extremely depressed, she told us.

Once again, pern pitched in with good advice. There were posters who had had personal experience with depression. There were professional therapists. There were other concerned citizens, or folks, if you will, offering good advice. And there were lots of hugs, beams, and solace for <z>.

> i feel such love for you in your dark trajectory, z, and offer no sound advice, only good will.

Then <z> posts that her daughter has just tried to commit suicide by swallowing thirty Tylenols! And almost succeeded! Wait. Was the daughter depressed, too? Must have missed something back there in the thread, but heck if I'm going to scroll back. This is costing money!

The groupmind is electrified by this news. And now <z> announces that her daughter "wants to speak to you directly." The daughter, who was now resting in bed at home, describes in rich, writerly detail her ordeal in the ER, the orange ID bracelet, the Band-Aid where the IV tube had been, the liquid charcoal she was forced to swallow in order to make her vomit. The post concluded with the daughter saying she "really love[s] the Internet" and thanking the group for supporting her mom.

At this point the average lurker, who knows nothing about any of the posters in the thread, rocks back from the screen, slaps himself on the head, and says Jesus Christ! For me this was strange because on the one hand this woman's life was absolutely falling down around her ears, but on the other hand I might as well have been watching a really good TV movie. This was entertainment! Then

<z> herself acknowledges the strangeness of situation by saying that it is kind of scary, "saying so much to cyberspace." But she reaffirms her love of this technology that lets her thoughts "flow out to intelligent, caring people" with the push of a button. It is a "priceless gift."

But it soon turns out that the groupmind is not quite the priceless gift <z> thinks it is. It may be a priceless gift in some ways, but it is capable of having the same dark thoughts about <z>'s motivation and veracity as any lurker. For if there was one thing the groupmind had no tolerance for, it was when a poster expressed untruths. Even the suspicion that the poster had "a secret agenda"—some motive that lay outside of the language of the postings—could set the groupmind off. The groupmind *was* easy to trick, and it knew it. It sometimes processed obviously incorrect data but couldn't tell the difference. It was prey to rumor, gossip, and superstition. In some respect the groupmind was not unlike a computer: very smart in some ways, very dumb in others. On the other hand, it was well aware of its failings, and frank about how easy it was to dupe. It relied on the truthfulness of the posters, and was quick to sense when something wasn't right.

So now one poster in the thread points out how strange it is that <z> is conducting this spectacle in public. Why is she doing this? Is this a way of avoiding dealing with her real life? Is this some sort of colossal ego trip? Is she making this up? She *is* a writer after all. Another poster points out that her daughter's writing style is strikingly like <z>'s, and then wonders aloud, Does she even have a daughter?

Like all metaphors, the commune metaphor—the idea of the society of mind, sitting around the fire circle, getting into each other's thing—could only take you so far on the WELL. It did not take account of the lurkers: the much larger group standing all around behind the fire circle, watching the society of mind, bored, hoping for some fireworks. Lurking was less like living on a commune than it was like watching TV, and especially for some of the younger pern, who were not old enough to have had the Sixties experience but *were* old enough to have stolen pot from their ex-hippie parents, watching TV was the controlling metaphor. The energy that the lurkers added to threads was hard to measure, since it existed only in the posters' idea of the lurkers. But the idea of the lurkers was powerful—you were never sure exactly how many people were out there in the darkness just beyond the fire circle, watching you.

The groupmind was well aware that the lurkers were out there, and it knew that even if <z> was innocent of playing to the gallery, it was entertainment, not empathy, that kept the lurkers coming back. And now the thread moved to this issue:

#334: pandora's vox (humdog)

i have read this topic with fascination—the same kind of fascination with which one watches a mouse being gutted by a cat, the same kind of fascination with which i listen to the words of jeffrey daumer and charles manson. i am not exaggerating for poetic effect, either. all of the images in this paragraph to me seem related to this topic—there are murders being committed here, for the amusement and entertainment of the locals.

first of all, this topic is one person's summation and opinions about a situation that involves more than one person. z has taken it upon herself to write her novel from the position of the Almighty—that is, she is the one who Knows what happened. furthermore, it appears from what she omits, and from the way she says things, that if you talked to some of the other Characters In The Novel, you might get a somewhat different vision of the events, even perhaps one that might seem somewhat more plausible than the one that we have been presented with here.

i see also that there is a whole bunch of people who have given assent to z's novel, and who want to be characters in her Book, who are willing to suspend disbelief, which is something that a Book requires, in order to participate in this particular spectacle.

i see also that a lot of language gets flung around here, lots of words that denote emotion, but that in reality, there is very little power, very little feeling being expressed. i note particularly the business about the daughter attempting suicide. this person who is alleged to have attempted to take her own life is not a cartoon character or an AI or a construct. it was a real person who did that, responding to a situation which sounds like a deeply weird and twisted one, but here, in this topic, the tragedy of a young girl has been exploited for its Entertainment Value, and what is worse is that the Exploiter is her own mother, and her mother has exploited the daughter for her own selfish pleasure and her own bliss; the bliss being that people tell her what a nice mommy she is, all evidence in earthspace to the contrary.

But now—another amazing bit of thread-vérité!—<z>'s daughter posts in defense of her mother, telling the unfortunate messengers of the dark side of the groupmind that they are wrong and cruel for saying what they say. How dare they accuse her mother of turning her tragedy into entertainment! Immediately following that post— you couldn't make this kind of stuff up!—another poster attacks <z> for putting her ailing daughter up to her defense, calling it "sick, tasteless, and dangerous."

After that the host of the conference freezes the topic, then un-freezes it, then <tnf> comes in saying what a train wreck this thread is, which galvanizes the pern to try at least to end it decently. A few weeks later <z> "scribbles" some of her postings—erases them from the thread, which was a privilege that posters on the WELL enjoyed—then later goes back and scribbles all of them. It was odd to go back to the thread long after it was all over and see her postings scribbled. From the groupmind's point of view, it seemed like a form of soul murder.

Why shouldn't <z> have been vouchsafed the same kind of transcendence, in her time of need, that the man in Woods Hole received? Because the groupmind didn't believe her, and it had believed the man in Woods Hole. And while it was possible to find out whether <z> had a daughter—and, in fact, the host of the conference did confirm she had a daughter—it was impossible for the groupmind to know whether she was writing her daughter's posts. And this to me was the point of the whole sad affair, a point that went to the heart of some of my experience on-line. The promise that the medium offered, which was the prospect of ecstatic union with the groupmind, or the Over-Soul, or "atman" (the Sanskrit word for the universal Self), or whatever you wanted to call that thing, was forever pegged to the freedom from ordinary responsibility and veracity that the Self in the real world was constrained by. And even if everyone did speak with complete truth and candor all the time, according to some sort of psychic honor system (which, of course, would never actually happen), the suspicion that some posters might not be who they said they were lurked around the corners of the medium, always there, a dark undertow.

IV. f2f ◊

As part of the Owner's expansion plans, the WELL had recently opened a New York office, and was trying to build a New York community, or "scene," as we say here. A young woman named Cherry Arnold was the WELL's pern in New York. She posted a message in the New York conference, announcing a potluck brunch for Howard Rheingold, who was coming to town on a book tour, at her place on Twenty-third Street. The WELL spirit kicked in; lots of people posted with enthusiastic offers of help, to which <hlr>, who was somewhere off in Nova Scotia, responded gratefully, and referred to everyone as "family." I wanted to meet Howard, but the posts about the yummy pumpkin cakes and extra-big quiches put me off, and also I was still pissed at <hlr> for his remarks about my writing in the Media Conference. In other words, I was a "flame ghoul." You are a flame ghoul when you feel you have been wronged, or at least had your delicate emotions bruised, in a thrash somewhere on the network, but you don't confront the perp there and then; instead you haunt the system, like Dante's Paolo and Francesca in Purgatory, except that rather than being summoned involuntarily by the word *"amore,"* you are summoned by the postings of your tormentor. You lurk in the thread as the pleasantries are exchanged between other posters, thinking to yourself, Oh, you're so friendly to everyone, aren't you? Well, it must be nice to be *such a big fat HYPOCRITE!!!* Finally you can't stand it anymore, and you appear in the thread, like the ghost of Hamlet's father on the parapet, and another thrash ensues.

At the same time, it felt good to be a part of things, good to give, and I'm sure I would have had a nice time with Howard Rheingold if we'd ever met IRL. I could also see how this odd leap of faith, of going to an f2f gathering of people who knew each other only from on-line, with no institution to select the group—anyone on the WELL who happened to see the posting could go to Cherry's party —and greeting each other as friends, was a positive, community-building thing to do. But I ended up not going to Cherry's brunch. I was thinking, What if I get there and I realize that all of this is merely the Nineties version of Jerry Rubin's networking parties in the 1980s? Then where would my notions of progress be?

A few months later Cherry started another topic in the New York Conference, announcing that she was throwing a WELL holiday party at her place, and this time I did go. My first f2f! An f2f is another curious on-line rite of passage, like a coming-out party for your body. But it is also the ultimate pledge of your faith in the power of virtual community, in a way. Do you really want to be friends with the pern IRL? Have their ASCII souls touched you deeply enough so that it doesn't matter what they are "really like"?

The basic dynamic of f2f was matching people's usernames to their real names. You said your real name first, then your screen name, and that was when the recognition came.

"Hi, I'm Michelle. On-line I'm chelbell."

"Oh, chelbell, nice to meet you!"

Then you tried to figure out what to do with your eyes. Eye contact was almost too intense, especially if you had been involved in some thrashlike circumstances with this pern now standing in front of you. Do you refer to your on-line wranglings? Or do you let that remain in the other world, running in background memory, and just make small talk, like you would with any other stranger at a party?

I saw one woman come in with an air of forcing herself to stay, linger nervously for about a minute with panicky looks around the room, then bolt.

At Cherry's I met John Perry Barlow. Barlow was a former Wyoming cattle rancher whose ranch had gone under in the late 1980s, and whose particular brand of "the would have" was to reinvent himself as a kind of cyberfrontiersman, an on-line version of Hoss, from the TV series "Bonanza." He was a co-founder of the Electronic Frontier Foundation, and a highly visible advocate for pioneer values on-line. He had a weakness for hyperbole, such as his statement that cyberspace is "likely prove the most transforming technological event since the capture of fire"; this nonsense came flying out of his mouth almost involuntarily, which made dialogue with him a bit difficult. He was wearing something around his neck that suggested a Wyoming cattle rancher's version of an ascot. I asked him where he lived, and he pointed at his head. "I'm Barlow at eff dot org. I've got two telephone answering machines, one in New York and one in Wyoming, and I got a cell phone and a modem. Otherwise I'm just up here." He tapped his skull again and looked up at me from under one eye, as though allowing me time to comprehend this mystical lifestyle. I suddenly wished we were not actually in "meatspace," but somewhere on-line, so that I could log out of this conversation easily. But I stayed with it and had a good talk about copyright, and the next day he e-mailed me an interesting piece he had written on the subject, arguing that the laws of intellectual property were made for the ownership of physical things, like books, and did not apply to the products of "Mind," which had no physical form.

The next day I logged on to the WELL to read the postings about the party, and saw a post from Barlow saying that one of the guests at the party last night had stolen Cherry's PowerBook. This was followed by a posting from <corrina>:

Topic 115 [ny]: NY WELL Holiday Party!! Saturday, December 10th, 9 pm #62: Michelle Waughtel (corrina)

This is positively disgusting. I don't suppose it is any comfort, but the culprit has surely fucked his/her Karma well into the next life not because of the monetary value of the object, but because of the cloud of mistrust which now must hang over our community here. I am deeply sorry such a thing has happened. Nonetheless, I would like to challenge us as a group to recognize this sort of thing as an invitation to let the light overcome the darkness . . . by remaining open and giving as before and not letting it discourage us to back down from the ideals that make this group so special.

V. **Neeeeeerrrrrrddddds!!!**◊ ◊ ◊ ◊ ◊ ◊ ◊ ◊ ◊ ◊ ◊ ◊ ◊ ◊ ◊ ◊ ◊ ◊ ◊

Back on the real WELL, a secret, private conference had been established by pern who were fed up with the Owner, and had decided, in the interests of preserving some of the qualities that made the *old* WELL so special, to found a *new* virtual community, to be known as "The River." When one of the hosts of the private conference sent me e-mail informing me of its existence, and saying I had been added to the "u" list, meaning I was empowered to enter, I was thrilled, though the fact that I was so pleased kind of appalled me. I went to the conference and saw the groupmind at its best— everyone doing their bit, all the posters patting each other on the back, all working in harmony, helping to organize The River. But instead of putting my shoulder to the wheel, too, like a good member of the community, or at least just lurking quietly, I immediately started to make trouble. In a topic entitled "The Dynamics of Power in a Cyberspace Community," I posted:

#160: My Time is a Piece of Wax (seabrook)

> But don't you think there is something ludicrous about convening a private conference mostly composed of older users to talk about the future of the well? baby, you are the past of the well. the future of the well is in the people you didn't let into this conference, probably because the subconscious reptilian part of your brain knew that it would be foolish to let the future of the well into a conference about the future of the well. fine. human nature. this world is like the real world, basically about protecting your patch of land. but at least let's have the decency to say it is that, and not that it is about the actual welfare of the well.

Reading this posting now, my eyes skitter away from my screen. One strange thing about living on-line is that so much of your life *is* preserved. Imagine a camcorder filming your whole life in all its glory and all its squalor, so that you and everyone else in your community were able to review all the pompous, mean-spirited, and embarrassing things that you once did, and that you otherwise might have mercifully forgotten.

#164: tiny speck/ruthless universe (peoples)

I'm unclear, seabrook. I sense a certain hostility on your part regarding those you regard as The Big Guns in this conference. I agree that I see some folks here as Big Guns and some not, but I bet most of that concept is of my own creation. I figure it's my problem to work out.

#167: kai (kaijohn)

The way I see it, things just aren't nearly so complicated or sinister. A lot of people have been attracted to the WELL because of what it is and what it isn't. Now the folks who have come together in a particular sort of place are facing a high probability that what they want to continue enjoying may not be possible any more.

Most of them will continue to use the WELL, but they (we) are taking reasonable steps to insure that there is a place in cyberspace for this kind of community, and this community in particular.

#180: Putana Happyface (mcintire)

Part of the Loki role is standing off to the side of the banquet hall, wrapped in a dark cloak and saying, "You're all full of shit. This is doomed to failure." I don't mind that so much, but the other part of the classical role is working to see that that prophecy comes true. I hope that you can be convinced that this is a much more egalitarian posting-ground and action zone than you had first assumed, seabrook.

Now I was back in archery class at Camp Wild Goose, in Maine. It was the summer of 1970. I was good at archery (it was another activity involving aim), but perhaps for the same reason I was also good at shooting sarcastic and half-nasty little barbs into the heart of this other kid in the archery class. Eventually the archery instructor, who was a Buddhist and a laid-back hippie dude from California—exactly the kind of person whom I now imagined I was tangling with in this conference—and who had already told me how appalled he was with my "negativity," got so disgusted with my wisecracks that he wouldn't allow me to speak. I had to keep silent. So I sat there kind of smirking, wondering how I could have turned out so rotten.

#299: David Dunning (dlee)

John, you almost invariably use language that sets you apart from some monolithic social body that you imagine is constituted by the ulist of this conference. You appear to have no wish to become a part of that group. For my part, you are welcome to be here, but if you feel so alienated from everyone here, can you say why you want to be here?

#309: Dan Levy (danlevy)

The WELL is the most powerful transference engine this side of one's own family.

Everybody projects onto their experience of the WELL a constellation of assumptions, personal history, ideologies, social theories and aesthetics.

So many thrashes have to do with stuff that has nothing to do with what it seems to have to do with, and this seems like one of them.

At the same time that my opinions were being decried in public, other pern were sending me private e-mail messages supporting me, which, of course, goaded me on to greater excesses of rhetoric and aggression in the thread. I was finally realizing that it was useless for me to think of the WELL as any kind of utopia—indeed, that was the source of my problems in adjusting to the WELL. It was like anywhere else, a place where there was authority to be grabbed, and where I would naturally figure out a way to grab it if I wanted to, just like I had in boarding school. It was ridiculous—another newbie mistake—to suppose that in the on-line world things were going to be any different than they are in the real world.

So, when <rbr> suggested that since I seemed to have such a fixation about the power of the hosts, maybe I should try co-hosting with him in Books Conference on the WELL, I was quite pleased —though I didn't act pleased. Actually I ignored him. When he repeated his offer in the thread, I sent him e-mail back, saying that I guessed being a host sounded kind of interesting. Within a month the WELL management had okayed my appointment and granted me comp time—all my bills were taken care of. I was permitted to put my username beside <rbr>'s on the Books Conference log-in banner, where everyone would see it when they came to the conference. And I felt good about all of this. I got almost as much attention being a host as I got being a Loki figure, and now that I was on the groupmind's side I didn't have to work as hard for it. I even began to like <rbr>. Maybe the pern wasn't so bad after all. . . .

Chapter Seven   Home on the Net

I. **Addicted to Bits** ◊

One of the great misunderstandings that the world has about believers in progress is that we are optimists. We may be cosmic optimists, but in our own minds we are as likely to be gloomy and pessimistic as cheerful. For many of us, the need to believe in progress is impelled by dread of the senseless tragedy, the freak accident, or some other terrible and bitter proof that the universe is arbitrary after all. Because we have no real faith, except in what can be proved by reason, we are especially vulnerable to such intimations.

Yes, you could try to listen to what the technology was telling you to do, but who really knew whether the technology was speaking for God or for the devil? Was it "good" to spend all this time on-line? Was it "good" that I now had to carry my laptop with me wherever I went? It was like a gun, in a way. I had willingly embraced the notion that at this very late stage of the second millennium, the path to spiritual enlightenment might lead through my screen, and as a result I had spent thousands of dollars in on-line and telephone charges, and thousands more on a desktop machine and color monitor, and devoted many hours to trying to improve my bandwidth and to downloading and configuring new and improved software—all the while thinking that these forms of material progress were going to enrich my life in some immaterial way. And where had it all brought me? To my room, sweating, junk piled around me, dim light, wires everywhere, eyes burning, neck aching, alone.

I had never gone in for any of the computer games that other people spent so much time playing on-line, but in fact, without realizing it, I had been playing a game of my own all along. It was like the war games Frenchy and I used to play, except that instead of weather, terrain, and troop strength, the variables were literature, history, and people. But the goal was the same: perfection through skill and practice; control that could be achieved in silence, and alone. Alas, poor Frenchy. Alas, poor me.

Aspects of life without parallel on-line had started to make no sense. The way mouths formed words, the gestures people made, their smiles, the light in their eyes—what was it for? Was it "good" that the journal in which I had been putting my thoughts about my on-line travels was now just a clippings file of my on-line postings? (Now, if I wanted to write something to *myself*, I wrote it in pencil on a big pad of drawing paper.) What had happened to the person I had been writing my thoughts to before, the person I had fallen into the habit of thinking of as "me"? That person seemed to be unavailable right now, and was perhaps blown away forever.

Certainly the times when I used to say to my screen "This is *great!!!!*" had faded. On-line was good for some things, but it was really bad for others. It was a good way to talk to strangers, but I regretted surrendering some of the quietness in my life to these voices, and I sometimes fantasized about logging off and never again logging on. In publishing your e-mail address, you are opening yourself to people, making yourself available to the world. It isn't quite like inviting people into your house, but you are fooling yourself if you think you aren't giving people a way into your life, in some way. There comes a day—after wading through fifty-eight pieces of mail on the mailing list you read, added to which are the two or three Usenet newsgroups, the WELL, ECHO, and other e-mail—when your correspondents overwhelm your capacity to respond to them thoughtfully, and the thoughtful part was what had interested me about e-mail in the first place. Ideas may be infinitely "expansible" throughout space, as Jefferson said they were, but time is not. That's why they call it "real time." It was hard to believe that Microsoft or anyone else was going to make a software robot ("bot") soon that I could trust to winnow the e-mail I wanted to answer from the e-mail I didn't. And anyway what was the point of having e-mail if you had to use a "smart" form letter to answer it? I drafted a letter from my bot:

> I am John Seabrook's "bot." I am a software entity that has accrued in John Seabrook's head, as a result of his receiving and answering so much e-mail. Am I a real or an artificial intelligence? It is impossible to say.
> Yours,
> John's Bot
> P.S. My bot can kick your bot's butt.

But I didn't feel right about sending it to anyone.

I wondered whether I was "addicted" to the Net. The metaphor of addiction is addictive (metaphorically speaking) when applied to on-line, because in some respects the technology *is* like a mind-altering drug. One is a "user" in both cases. It loosens the tongue, relaxes the inhibitions, makes it possible for you to be funnier or wiser or more charming than you are "straight" (i.e., without the screen). You love it, but you also hate it. You crave it with the same urgency with which one craves just one more cigarette. The only ontological difference between Chat and getting stoned is that you don't smoke the bits. (And also on-line is more expensive than pot. . . .)

On the other hand, the average American watches something like five hours of TV a day, and nobody calls *that* an addiction.

Naturally, I turned to my screen for more information about net dot addiction. Somewhere on-line I had heard about a mailing list called the Internet Addiction Support Group, which was started by a New York–based psychologist named Ivan Goldberg, who had a practice in what he had dubbed IAD—Internet Addiction Disorder. So I subscribed to the list, and a "bot" sent me a message back which said,

From: <listserv@netcom.com>

Welcome to the Internet Addiction Support Group (IASG). This group was created to help people who consider themselves addicted to Internet or other on-line services.

Having a group devoted to Internet Addiction on Internet may be like AA holding meetings in a bar. But, there seems to be no other way for a group such as this to get started.

I lurked for a while. Most of the posters to the list seemed happy about their addiction, if that's what it was, and showed little or no desire to be cured.

From: "Janet Williams" <williams@eliza.netaxis.com>

Find myself at the computer at least twice a day at home and for eight hours a day at work as well. Am amazed at the variety of subjects of interest available on the Internet and find something new and interesting each time I am on-line. Send e-mail regularly to people I have never met in both Virginia and Australia. Guess you can say I am hooked. Still get a thrill out of getting e-mail!! Heaven forbid if I ever lose my electricity in a thunderstorm. Bet there are others who feel the same way. . . . Janet

Even people who did regret the time they spent on the Net seemed scarcely able to contain their pleasure in their condition, like a doctor in Florida who wrote of on-line,

> slurping up thousands of life's seconds like a voracious anteater in a giant colony. My fingers dance on buttons and I can feel my time on earth being shortened, my vitality being sucked, my head spinning. I'm using these fragile moments of our brief vanishing years, these precious minutes of lucidity that crumble sooner than we think, not to answer human correspondence, not to record my thoughts, not to do good in the world . . . more moments of my life gasp like guppies and flop over gone. And I can't help it. I can't stop.

Some posters even wrote inspirational postings about how Net addiction had improved their lives:

From: <ids@mail.internet.com.mx> (ids)

Hi everybody!

My name is Pedro Hernandez, I live in Mexico City, I am 36 years old. I have been working in software development and computer training for the last eight years.

I discovered Internet on last december when a Internet service provider moved their sales office to the same building where I have my office.

I had the opportunity to search in the Internet for free for a couple of months and that was the way I became Addict. I set four time records of use of my account from February to June. On June I was surfing over 100 hours.

Around April I began to stay at home playing on the Internet instead going to work. Fortunately I am the owner but some of my customers began to complain of my absence at the office.

Finally I decided: If the Internet affects your job, Quit the Job!

I made some arrangements with my employees so they are running the business at their own risk and I decided to become an Internet Trainer and Consultant.

Today I am running Internet Seminars with a great success. Even some companies as American Management Association is preparing an Internet seminar where I am going to be one of the main speakers.

Also I am preparing the first Spanish Version of a Virtual Magazine that will appear on September. If any of the addicts wants this monthly magazine please send me a message and I will add your e-mail to the subscriber's list.

Sometimes you may turn an addiction into a good job. And no body complains about your 12 hours sessions on the computer.

My final message is: Keep surfing and search for living of The Net.

Pedro Hernandez

Yes, I could tell a few inspirational stories about the Net, too. For example, take the story of my brother-in-law, Michael Neubarth. Like me, Michael was an English major in college. He earned a master's degree in English literature from Columbia University, where he wrote his thesis on the influence of the German Christian mystic Jacob Boehme on James Joyce. Michael also played bass guitar in the Royal Appliances, a band that won the Battle of the Bands in the My Way Lounge in Elizabeth, New Jersey, in 1983, defeating the Young Turks, John Wayne, and the Cucumbers. Like me, he got into computers by accident. When I first knew him he had a job writing for a weekly newspaper that went out to people in M.I.S.—management information systems. I knew his job was related to computers, but that was all. I remember him asking me once when our wives the sisters Reed got together, "Do you even know what I do?" And in fact I did not know what he did. It was just something related to computers.

Then in 1993 Michael got a job as the editor-in-chief of *Internet World,* a year-old magazine started by the MecklerMedia Corp., and around the same time I sent e-mail to Bill Gates, and before long we had a million things to talk about when we got together. Moreover, our wives were more or less forced to take us more seriously. After all, we had been clever enough to latch onto the Internet just before the masses did. We had even been asked to talk about the Internet on television! The circulation of *Internet World* had increased fifteenfold since Michael took it over. The man was a genius!

But not all the posters to the i-a-s-g list had such inspirational stories to share:

Dear Group,

I am a 36-year-old-male who is dealing with the demise of my marriage due to my wife's on-line addiction with a D/S chat group on Prodigy. I removed the modem from our home computer when she announced her intentions for divorce. She stated that I was too controlling in asking her to modify the amount of time on-line that she was spending so that we could have more time together. I should also state that this occurred after I was stationed away from home for 13 months in Guantanamo Bay, Cuba, with the Navy, and she developed her on-line activities as a support mechanism. I am a doctor currently serving at a naval hospital. I would appreciate hearing from anyone else whose on-line activities or activities of their loved ones have had an adverse impact on their life.

From: <anygal@camden.ge.com>

Well, for me, on-line romances took away from my marriage in that I was obsessing about the person I was involved with on-line and investing all my emotional energy in THAT relationship rather than the REAL relationship with my husband. And a mere mortal husband could not compare with the fantasy I built up about this *unknown* person.

I quit out of two on-line romances (both were with people within driving distance) when we started discussing meeting real-time. I do not wish to jeopardize my marriage; that is why I joined this group. I can't go into chat rooms anymore because I always hook up with someone and it always follows the same path . . . I become totally obsessed. Maybe some people can handle these little romances as *harmless fun*, but as a person with an addictive personality, I cannot.

That was a good measure of addiction: when it began to interfere with your relationships with people you loved IRL. Was this a problem for me? I wouldn't say on-line seriously interfered with my marriage, certainly no more than my compulsive work habits did, although on-line in particular did seem to annoy Lisa from time to time. For example, one evening around this time, as we were walking to have dinner in the neighborhood, I said, "Someone made a really good post today on the WELL."

Lisa responded, "What did they say? *'Oh, John, you're so smart, I agree with you so completely'?*"

Later, however, when I rescued some data she thought she had deleted on her PowerBook, I was "my brilliant husband."

A mother posted a troubling account of her teenage son's net dot addiction:

My son was spending over 12 hrs a day on-line. He began skipping school, lying about being on-line, and he hacked my password to attempt to be on-line without detection. (Luckily Mom's had twelve years on the Internet and still had a few tricks up her sleeve!)

My son was hospitalized for depression as a result of trying to pull himself off of IRC chats and MUDs. He spent 30 days in a hospital that treated him just like he was addicted to alcohol or drugs. (He fit most of the symptoms!) My son has stayed off of computers except to use them to write school papers (and at first he was supervised doing that). So far he hasn't gotten back on the Internet. He has made up all of his school work from last year, doing well at school this year, working part-time (20 hrs/ wk), and continues in therapy.

Quite a few of the posters were teenagers:

From: <BritSkye@aol.com>

hi everyone!

my name is Britta, i'm 17 (well, i will be in 7 days!), and i'm new to this group. . . . but i am DEFINITELY an online addict!!!!!!!!!!!!! it has gotten way out of control. my last two bills were $200 each, and for an unemployed teenager, that is total torture!

i go online whenver i can, for hours and hours at a time, and my whole family complains that i'm "gone" . . . i'm not the same person; they hardly know me anymore. . . . i get so absorbed in the online world that i tend to neglect, and not even see, the real world. for instance, my parents and brother had a 20 minute long

SCREAMING match just one room away from me, and here on the computer, i wasn't even aware of it. at all. i was totally deaf. . . .

when i go on vacation, no matter how fun the trip is, i miss AOL terribly and write emails in my head to be sent when i get back . . .

i never eat breakfast or lunch anymore, or even get up to pee, cuz i'm always online until dinner. . . .

i have signed on at 10 pm and not gotten off till i've seen the sun rise. . . . .

my friends feel neglected, and have even sent me "hate mail", saying i'm "never around" and "ignore them" and "why don't i ever call?"

i was ONLINE on CHRISTMAS DAY. . . . dad: "Brit, you ready to open presents?"

me: "yeah, hold on. i just have one message board left to read." dad: "brit? you done?" me: "yeah, almost. i'll be right there . . . " dad: "Britta. it's *christmas*, shut OFF the computer!" me: "yeah, yeah . . ."

SO?!? do i qualify for this group?????

Dr. Goldberg himself posted his thoughts about "Internet Addiction Disorder" to the list:

This may make me persona non grata around here, but here is my take on IAD.

Computers and the Internet are no more addicting than work. IAD is very much the same as workaholism. Despite the (poor) name I gave it, there is no true addiction involved in IAD just as there is no true addiction to work.

People overwork or overuse computers/Internet because they do not want to face some unpleasant reality about themselves, their significant others, or their situation in life. Some of the major things that I have seen people try to avoid confronting by engaging in overuse of computers/Internet have included:

lack of social skills

lack of friends

depression

unfaithfulness of their SO

drug abuse of their kids

lack of meaningful plans for the future

fear of working hard at school or a job (fear of failing despite hard work)

unconscious self-destructive impulses

I am sure that this list is incomplete, and many other reasons for computer/Internet overuse can be found.

Finally, I introduced myself with a posting:

From: <seabrook@echonyc.com>

Hi everyone! I'm new to the group.

Last week we gave a dinner party—eight people altogether, my wife, my sister-in-law Terry, whose birthday we were celebrating, and three friends. In the middle of dinner I had a sudden powerful urge to leave the table, go down the hall and check my e-mail. I figured it would take about eight minutes, then I would come back to the table and have the new e-mail running in my background memory while my attention was focused on what was going on at the table. My feet started turning under the table, computerward. Was I bored with my family and friends? No. Did I want to be alone? No. Was there something I could get from e-mail I couldn't get from the conversation at the table? Mmmmmaybe. What was it? I managed to stay put until after the elevator door had closed on the last of our guests, then I immediately went to the computer and logged on, and felt the familiar jolt.

After I had posted that, I logged on eagerly over the next couple of days to see if anyone had responded to my posting, but no one did. Hurt by the group's apparent lack of interest in what I regarded as a nice little aperçu of the virtual life, I slunk back into a lurk. And in doing so I realized that I wasn't primarily interested in supporting others in the group, but in getting attention for myself. I wondered how many other members of the group shared my motive.

As was true of many of the mailing lists, the i-a-s-g list was easier to subscribe to than to unsubscribe from. Some people signed up, thinking the group sounded neat, and then discovered each time they logged on they had ten messages waiting that would take another two minutes to download. But in their jones they had neglected to save the instructions about how to unsubscribe.

From: <lodev@innet.be> (Lode Vermeiren)

Subject: I AM GETTING SICK OF THIS!!!
    I already send a dozen of UNSUBSCRIBE messages, but the only thing listserv@netcom.com says is: lodev@innet.be is not a member of i-a-s-g, but i still receive messages sent to the list.
I WANT OFF THIS LIST !!!
PLEASE HELP ME!

The only thrashlike event in my time on the list was provoked by a poster named Lorraine Day, who had recovered from real-life addictions to both alcohol and drugs, and now was apparently addicted to the language and feeling of recovery. She sent the list mail three times a day—Morning Thoughts, Evening Thoughts, and Points to Ponder, filling our mailboxes with sunshine. I hated it, and found myself flying into bizarre rages at Lorraine. When I calmed down I wondered whether this was a symptom of my Internet addiction—it angered me to have someone like Lorraine treating it like a real-life addiction. Eventually Dr. Goldberg posted a message saying he thought Lorraine's postings about recovery were "off topic," and asked her to desist.

From: "Ivan Goldberg, MD" <psydoc@netcom.com>

This is a formal request that you stop posting off-topic material to this list. If you choose to ignore this notice you will be immediately unsubscribed and prevented from resubscribing.

This post naturally produced posts defending Lorraine's right to speak:

From: <luckydog@mars.superlink.net>

Dear Ivan,

Stick it where the sun does not shine. Leave the lady alone.

Regards,

Dean

I even emerged briefly from my lurk to make a posting.

From: <seabrook@echonyc.com>

Evening Thoughts: Should I change the cat litter?
Points to Ponder: I will try to be less sarcastic.

When Lorraine wouldn't stop making her inspirational postings, Dr. Goldberg apparently "kevorked" her—another useful onlineism, which I had picked up on ECHO, meaning "terminated."

II. **The Return of Baal: A Jeremiad** ◊ ◊ ◊ ◊ ◊ ◊ ◊ ◊ ◊ ◊ ◊ ◊ ◊ ◊

I also searched for an answer to the question "Is this good or bad?" on the Web. By now I had developed a new Web surf technique. Rather than just sitting there and asking myself, "Okay, what do I want to know?" I tried instead to relax my conscious brain and to snorkel over the underwater reeflike parts of my mind. The associative nature of the Web seemed more compatible with my subconscious mind, and giving in to that side seemed to make for a more satisfying Web experience.

Out of my relaxed mind a word appeared—and the word was "Baal." Yes, in fact, I had been meaning to find out more information about the ancient god Baal. For years now some cult anthropologists, who believe that ancient Baal-worshipping Celts sailed from Europe to North America about twenty-eight hundred years ago, had been coming to my parents' home in Vermont to look at their root cellar, which is next to the house. The believers say that the root cellar was built by these ancient visitors as a temple to worship Baal, and that you can see the word "Baal," written in the ancient language of "vowelless Ogam," on the lintel. These people did not appear to be Baal worshippers themselves (although, of course, one never really knows); many were simply elderly WASP retirees trying to keep active with their hobby. However, having all these pilgrims appear on our property to marvel at our root cellar was something of a nuisance. When the retired Episcopal reverend who lives just up the hill from us was asked by strangers if he happened to know where the "site" or the "Baal chamber" was, he liked to reply, "Oh, you mean the root cellar!," which sometimes caused people to chuckle knowingly and say, "Oh, yes, yes, *the root cellar.*"

The notion that an ancient, Baal-worshipping people built our root cellar seemed pretty nutty to me, but I found that, once lodged in the noggin, the idea wouldn't come out. In planning cross-country ski outings I'd find myself saying to Lisa, "Let's go by the Bride of Baal's chamber today," that being a second "site," on the other side of the hill. One night, poking at the fire, Lisa said, "Hey, John, look! The face of Baal!" And there the old god was, in the burning logs, with a short, curled beard of ash. My brother-in-law Michael suggested I put into my e-mail "signature file," which was the quotation box that the Eudora program allowed you to attach to your outgoing e-mail, the following made-up bit of Baalistry:

"I danced like a puppet, like a minion of Baal."

I did put it in there, thinking it might coax a few Baal worshippers out of the woodwork, but not a single one of my Net correspondents even commented on it. Perhaps everyone just assumed it was true.

At any rate, I had Baal on the brain, so into the little window of the Infoseek "search engine" I typed Baal and then hit <return>. By the beard of Baal! Over one hundred hits! The first site was "Canaanite/Ugaritic Mythology FAQ, ver. 1.0":

Baal (also called Baal-Zephon (Saphon), Hadad, Pidar and Rapiu (Rapha?)—'the shade'), the son of El, the god of fertility, 'rider of the clouds,' and god of lightning and thunder. He is 'the Prince, the lord of earth,' 'the mightiest of warriors,' 'lord of the sky and the earth' (Alalakh). He has a palace on Mt. Zephon. He has a feud with Yam. His voice is thunder, his ship is a snow bearing cloud. He is known as Rapiu during his summer stay in the underworld.

The second site was the home page of "Stace A. Baal":

Hey folks! Welcome to my little exit off of the information highway.

I am currently working for the University of Southern California, School of Medicine, Instructional Imaging Center as the Imaging/Interactive Specialist. I also attend classes part time working toward my communications degree. When I'm not at work or school I'm usually outside trying to enjoy myself in the California sun.

At the "House of Yahweh Home Page" (http://yahweh.com/), I read scripture from the Old Testament concerning the covenant God made with the children of Abraham, in which they promised not to worship Baal:

Deuteronomy 8:19–20

19. If you, by any means, ever forget Yahweh your Father, by following hinder Gods (Elohim) to serve and worship them, I testify and witness against you this day that you will surely perish.

But the children of Yahweh broke their covenant with him:

Judges 2:13

They provoked Yahweh to anger, because they forsook Him and worshiped Baal the Lord, and the Ashtoreth; the goddess Astarte or Easter.

And they did indeed perish.

At "Bob Baal's Webpage" I saw:

"Here is a JPEG of me. It's pretty horrible—YOU HAVE BEEN WARNED!"

In an on-line essay by a born-again Christian economics professor I learned that Baal was not just one god, but the name given to the presiding deity of any given locality. Baalim (the plural) owned the land along with all that it produced: crops, fruits, cattle, etc.; in worshipping that local god, you got to own the property. The professor went on to argue that people who rely on the secular institution of civic government, separated from the teaching of the church, are practicing a modern form of Baal worship:

> They all appear to be worshipping at the altar of an idol, the idol of modern humanistic civil government. In short, they might well be regarded as modern Baal worshippers.

About sixty hits deep into the search, at the Web site of Jay Kinney—by coincidence another *Whole Earth Review* writer (if there was such a thing as coincidence on the Web)—I came across an article called "The Return of Baal." Kinney saw Baal as a kind of Jungian archetype, an image from the collective unconscious, and linked the return of Baal to the rise of heartlessness, oppression, warmongering, and spiritual corruption caused by capitalism.

> How far one wants to pursue this line of inquiry may depend on how far one is willing to associate Baal with evil and project that value onto modern America and capitalism . . .

While on Kinney's home page I clicked onto another one of his articles, in which I came across these remarks by Lawrence Wilkinson, co-founder of the Global Business Network:

> Just as during the Enlightenment 'the nation-state' took over from 'the church' to become the dominant seat of action, so the nation-state is now receding, yielding center stage to 'the marketplace'; the action in the marketplace is interestingly everywhere:

local, global, wherever—where 'wherever' is increasingly dictated by 'pure' economics and interests, not by national borders (nor the tariffs, national practices, and customs houses that define them) . . . I believe that we're in for some nationalist noise and some nationalist violence before the transition is done, but I do believe that it will finish, to be replaced by the kinds of tribal and commercial conflicts described by folks like Joel Kotkin *(Tribes)* and Charles Hampden-Turner and Fons Trompenaur *(Seven Cultures of Capitalism)*. What will remain of nationalism? My bet is that it'll have the character—the strength and relative 'weight'—of brand loyalty; perhaps in some cases, that charged variety of brand loyalty, a fan's relationship to a sports team.

That surf session, with little surflets off into other Baal-related material, had lasted about two hours. I had been sitting at my screen as though before an altar—I'm sure that if those ancient Baal-worshipping people in our root cellar could have seen me, they would have assumed I was worshipping *my* God. The muscles in the back of my neck were tight and aching. However, I was pleased with my surf. I had no idea what to make of all the random bits of information I had accumulated, but it felt as if somewhere in my mind the pieces were chunking down into a pattern, down at the level where the Jungian archetypes live, where the face of Baal that Lisa saw in the fire blended in with the mandala image from Jay Kinney's Web site, the scraps of leaves and earth in the root cellar, the idea of all of us belonging to brand-states like Nike and Disney, and my fear that in giving ourselves over to this technology we will all somehow end up worshipping Baal.

I surfed over to the Internet Society's Web site and read an article by Vinton Cerf saying that 1994 was the year business discovered the Internet. "That's true," I said to my screen. I remember the first time I felt the commercial interest, during a thaw in January 1994, when I was jogging along the Hudson River promenade in lower Manhattan, wearing a T-shirt that my brother-in-law Michael had given to me. It had the word INTERNET printed in big red letters on the front. As I ran past the businessmen strolling around outside the World Financial Center, smoking their cigars, I noticed they were looking at my chest with greater interest than at the T-shirts saying MICHIGAN STATE, or THE BLACK DOG, or SEX WAX, IT'S GOOD FOR YOUR STICK. Six months ago they would hardly have noticed, but now their moneymaking instincts were apparently aroused. A feeling not unlike pride made my chest expand a little and my legs get springier, and as I jogged past them I started thinking to myself, my thoughts coming in the rhythm of my breathing, "I am team INTERNET, I am infinite, I am always growing, I am a one-word version of 'I'd like to teach the world to sing,' I am a virus for good!"

A few months later I logged onto CompuServe and saw a banner that said, "Spend Easter in the Electronic Mall!" Then "spam" celebrating the wonderful business opportunities offered by the Internet began coming to my various mailing lists and e-mail addresses, in ever-growing quantities, like Chinese menus stuffed under my virtual door.

Imagine starting a business that requires:

- No Capital Investment
- No Products to Purchase
- No Inventory to Sell
- No Delivery of any Product
- No Collections
- No Quotas
- No Employees
- No Customer Risk
- No Experience

It's absolutely the most perfect business opportunity ever!

Then I started hearing people at parties talking about AOL's cash flow, and how the key to making a succesful on-line product was to brand it. Then Web site URL's started showing up on ads for the Super Bowl, on the sides of Zima beer bottles, and on graffiti scrawled on a wall in Prince Street, in Soho. Then corporations started registering domains for their products—metamucil.com, diarrhea.com, pimples.com, underarms.com, badbreath.com, freshness.com, toiletpaper.com, americascheeseexperts.com, saladdressing.com, and parkay.com—producing a glut of soul-wearying garbage equal to anything that the hero of *Pilgrim's Progress* encounters in Vanity Fair. Then bad programming that the public had learned not to tolerate on TV came to the Web, repackaged as "interactive." It was still junk, but now it was only a mouse click away! Then, in an effort to figure out how to combine the "real time" component that makes television compelling with the interactive component of the Web, companies started churning out garbage that was both bad television programming *and* bad Web content. I went to a cybercafe on St. Mark's Place for the taping of one of these events: a "live" debate, sponsored by Microsoft, that was supposed to be fed to its Encarta Web site, so that surfers sitting before their machines at 9:00 P.M. EST would see it, just like, say, "Seinfeld." The television production was cheesy looking, like something from the very early days of cable, and the Web component didn't seem to be working; and anyway, the last thing one wants to do on the Web is wait for some "live" programming to be uploaded. As I sat at one of the terminals in the back of the room, waiting for the show to come on the Web, I felt the hot lights of one of the circling local news crews that had turned out for the "event" come up on my back.

"Don't look at the camera," the producer said. "Just, you know, surf or something."

But the Web site still wasn't working.

I said, "Do you realize what we're doing here is taking bad television, making it worse by putting it on the Web—which isn't even working right now—and now you're turning *that* into programming for the evening news?"

"I know, don'cha hate it?" the producer said.

A young woman who was helping to promote the event looked at me with concern. "I think you need to talk to somebody," she said.

"No, please, I promise I'll keep quiet from now on."

"Nooo, I really think that you're just not getting what's so *special* about this technology. Richard, could you come over here?" Richard, an enthusiastic young Microsoft cybersoldier, appeared and began to say, "Well, the great thing about this technology is it's entirely *interactive*. You get to see what *you* want to see, *when*ever you want—"

"No, please, stop. Please. I mean, I understand all that. I use it. It's just that—I'm sorry, I don't mean to get upset . . ." But as usual, I did get kind of upset. I never should have left my room.

Although the Web contained an interactive component—you still had to ask for the things—it seemed to be the end of interactivity in other ways. The Web threw the leading edge of the technology back toward broadcast, back toward the one-to-many; the Web represented the triumph of the lurkers over the posters. True, in theory anybody could have a channel and be a broadcaster. But in an information society such as the Web, all the members have to have their nuggets of information, and the poorest had only charcoal to set in front of their corrugated-tin huts, while the richest had glittering and irresistible palaces of mind candy. Even though the www.tampax.com was the same number of mouse clicks away from my computer as a high school kid's personal home page, Tampax had the kind of costly programming that only big corporations can afford, and the superior graphics and other bells and whistles effectively wiped out the democratizing potential of a distributed network. And since no one had yet figured out any way of making much money from Web sites, only corporations with large promotion and marketing budgets could afford to build expensive ones.

I made an effort to avoid the corporate sites, and to surf the "personal home pages," on the Web, the pages that were put up not for commerce but for community, or at least for noncommercial self-promotion. A personal home page was like an e-mail address, a place on the Net where people could find you; but whereas an e-mail address was just a mailbox, a home page was a front porch. You wanted guests to have a good time when they visited your home page, and you hoped they would take away a favorable impression of you. You could link your home page to the home pages of friends or family, or to your employer's Web site, or to any other site you liked, creating a kind of neighborhood for yourself. Because a lot of the people putting up personal home pages were college students, their pages tended to resemble college dorm rooms, with pictures of Bono, their class schedules, and so forth, but in theory you could design your page in any way you wish, and furnish it with anything that could be digitized—your ideas, your voice, your causes, pictures of your scars or your pets or your ancestors.

At one home page, I saw the list of the contents of a purse that the guy's ex-girlfriend had left at his home.

Vanity mirror (6x10 cm)
Lifestyle condom (ribbed for HER pleasure)
A penny

A financial responsibility card for her Sunbird

Citgo Plus Credit Card

Kaufman's credit card

Her ATM card (damn, I used to know the PIN too.)

Sprint calling card

Half of a Military I.D. card from one of her many lusts.

American Red Cross "Community First Aid" certificate

American Red Cross "Community CPR" Certificate

American Red Cross "Blood Donor" card (A +)

YMCA PTC for lifeguarding

Certified lifeguard patch from the YMCA

17 credit card receipts totaling over $680 (and not a penny on
me!)

Business Card from Dan Michaels, owner of Clarion Western
Auto

Sterling Optical business card

ATM receipt for $5, and a $22.96 balance

Her checkbook with one check left

On the home page known as "LoserNet" you could read the
writer's serial account of the activities of his upstairs neighbor:

Friday

Loser woke up early as usual and took his twenty-minute
shower. Then silence. The slow, laborious clockwise pacing
began and continued until he had to leave for classes. A short
phone call was made as is done every morning. Either he was
calling "stooge" or he was checking in with his mommy. I left
and did not get in until after 11 P.M. I thought that Loser might
have gone out somewhere. Within a few minutes Loser got up
and started pacing around. It is Friday night, for goodness sakes!
Finally, he either sits down or goes to bed. Nope, he's back up
again. He has to take a whiz.

If you wanted people to "hit" your site, it helped to have appealing content to offer. The simplest way to do this was to make a "hotlist"—a collection of clickable links to other Web sites that you thought were interesting. The mind-boggling speed at which sites were being created, combined with the lack of order on the Web, meant that keeping any kind of authoritative watch on Webalicious sites required a great deal of surfing. Therefore many Websters were willing to surf for you, in return for which they got you to hit their site.

The basic idea of the hotlist was refined over and over across the Web. For example, any woman who put a personal page on the Web was liable to find herself listed on Robert Toups's "Babes on the Web" page, which was one of the more controversial Web sites. Women were assigned a rating on his Toupsie Scale, from one to four. It was the basic idea of the hotlist, with value added by a resourceful information entrepreneur, Mr. Toups. On his site Toups wrote:

> Placing a Home Page on the World Wide Web is an invitation for entry. Having a personal photo on that page is an invitation for it to be rated based on the TOUPSIE SCALE.

Women who didn't want to be linked to "Babes on the Web" could e-mail Toups and ask to be removed, though he didn't make any promises. I saw that Shawna Benson, one of Toups's "Babes," had written on her home page:

> I am now listed in Rob Toups's lovely "Babes on the Web" page! When I first saw this page, I wasn't sure if it was good or bad. I have since decided that it isn't too bad—considering the added traffic my page now gets! So check out his brilliantly designed page.

and had added a clickable link. (I also saw, at the end of her "Read All About Me!" page:

Oh and last but not least, I love Disney. I don't care if it's this huge empire that will take over the world, I love, love, love Disney.)

There were hotlists dressed up to look like tipster sheets. One of the most successful of these was called "Suck":

> At Suck, we abide by the principle which dictates that somebody will always position himself or herself to systematically harvest anything of value in this world for the sake of money, power and/ or ego-fulfillment.
> We aim to be that somebody.

Imagining these Web-savvy youngsters going about their work, I thought of the Tony Curtis character in *Sweet Smell of Success*. These new media creatures appeared just as the last of the press agent–journalist go-betweens died out. . . .

There was obviously an element of self-promotion involved in putting a personal home page on the Web. Someone sent this anonymous posting about personal home pages to one of my mailboxes:

> The thrill of "making a home page" is tied to the same impulse by which one practices his signature on the back of an envelope. Similarly, as one refines his penstrokes—takes in a curlicue, slants his L's—we now modify, sculpt, and massage our home-pages. The 3W stretches that idea, though. Where once only the writer could see that signature, it is now publicly broadcast. We are ham operators with microprocessors. And there are millions of us.
> But why? What drives someone to construct his own page, detailing, literally, to the world, the state of his girlfriend, his automobile, his interest in antique toy collecting?

The Homepage has been called the Canvas of the Individual, and we have chosen to obsess over its content because we know no better satisfaction than the dissemination of our own tired, trivia-obsessed selves. At last, others might notice how one changed the slant of his signature.

But putting up a home page was also an act of joining the community of the Web, by sharing what you hoped was useful information with others in the group. The point at which the self-promoting ended and the public-spiritedness began was very hard to place. At any rate, one sometimes heard the voice of youthful idealism on the Web, which was harder to find in print and on TV, where young voices were so thoroughly marinated in irony. The most celebrated callow youth on the Web was probably Justin Hall, a twenty-year-old junior at Swarthmore, whose "Justin's Links from the Underground" was a famous site. Interspersed with links to racy sites were Justin's own writings, which he also sent to people on his mailing list, and for which he solicited donations, like an on-line subway busker. Among the odds and ends found under "Justin's Writings" was this cyber-rap:

Go out into your neighborhood and do a video documentary!

Stage a play on a street corner! Strike up a conversation! Read a poem on a train!

Then, write about it on your Web page.

Remember the first time you had sex? How strange that was? Write about it. Put it online.

Remember the first time you were dumped? How shitty that was? Write about it. Put it online.

I'd sooner read that than Barry Diller's five means of media ascension.

Culture doesn't come from Warner Brothers and Sony. Culture

is that woman friend of yours who tells the most outrageous stories.

Culture doesn't cost big bucks, and hang in a gallery of modern art. Culture is your friend who likes to draw. . . .

The web is an opportunity to make good our fifteen megabytes of fame.

Still, idealism was the exception on the Web; on the Net it had been the rule. Money and marketing were the rule on the Web. The communicators had come first; now the traders were arriving, having discovered that the power to market was made great by the same technology that made the power to communicate great. The word "community" was easily forged into invitations to join the "Molson community" or the "Frito-Lay community" by forking over some information about yourself, in exchange for which you would be allowed into the Molson clubhouse and be eligible for all sorts of fun stuff and free prizes and other kinds of "recognition." It was all about recognition. At the Kellogg's Web site, kids could get e-mail from Snap, Crackle, and Pop. At least when kids watched television they knew when they were watching a commercial and when they were watching a cartoon. On the Web, it was a lot harder to tell the difference. Maybe there was no difference.

I had been thinking I was retracing Christian's steps in *Pilgrim's Progress,* when in reality I was moving my little thimble around the Monopoly board. That same elevation of the spirit, which I had felt almost two years earlier on first getting into e-mail, that special lightness of hope and possibility which caused people to look up from their screens and yelp "This is great!," presented a first-class marketing opportunity. Look, a little window into the consumer's soul! What a nice surprise! And here we have been going to all this trouble to come up with characters like Joe Camel, cigarette spokesbeast, in order to get this kind of purchase on young consumer! We'll just slip through this little window and market these young souls into oblivion!

In spite of the utopian promises made by the promoters of the Net, I didn't notice traditional media powers getting any weaker. On the contrary: Instead of distributing power to the edges of society, the Net offered the media megamachine a new way of consolidating its hold. The Net would not develop into a revolutionary new medium that replaced existing media—the people who used that kind of rhetoric (like me, in my newbie days) were like fog machines. They obscured the truth. What was more likely to happen, it now seemed to me, was that the few advantages and innovations that the Net offered would be seized by the megamachine and used to further entrench itself in our daily lives. And with the growth of the corporate Web sites, it appeared that one of those innovations was a new way of marketing off-line goods and content. Net dot marketing got into your head in the same way that, say, MTV got into your head —it worked the brand and the desire to have it right into your cortex, like the mink oil I was forever massaging into my leather boots, to soften them.

What if the Web was an egg laid by the military-industrial complex that had taken forty years to hatch? The utopians had swallowed the egg, and now, full of cheerful American enthusiasm for social progress through technology, were in fact unwittingly working to accomplish the opposite—helping to lay the wires for the greatest trading and marketing machine that the world had ever seen. The Net was media, after all. It was the buzz. It wasn't like people were getting land, or cattle, or better housing. They were getting home pages. Did we really think this was going to make people happier, wealthier, more efficient? With all these distractions that did not exist before, these tempting media pleasures? "Distracted from distraction by distraction" is a line from T. S. Eliot's *Four Quartets* that applies nicely to some aspects of the on-line world.

III. E-mail from Dad ◊

Hmmmm. . . . What *would* it be like to have a Web site of my own? I imagined it would be like starting a restaurant, except that instead of serving food I'd be serving information and recognition. A place where, if you found one of my postings elsewhere on the Net interesting or useful, you could come and find out how to get more of the stuff. (Honey, look at this man's information! Ohhhh, let's get some to take home to the kids!) Maybe the site could also be a place to republish work that had grown and changed as its author had changed. When you are published in print there are no opportunities to dial into the server late one night, enter your password, and alter your book or magazine article, silently, as it sits there on the bedside tables of your readers. What I had in mind was nothing so elaborate as www.tampax.com, just a simple cabin twelve logs high, with improvements to follow. Something with words for walls, scanned-in pictures for decoration, maybe some hotlinks to the places where I published my stuff in print, and an e-mail button.

I wasn't thinking of *charging* people to visit my site. (No one would have paid, for one thing.) Building a Web site was more a way of casting certain on-line experiences into the form of "transactions," or "hits," which were the basic units of currency in the transaction economy (a word that nicely captured the odd duality of the experience, which was both aggressive and seductive, both selfish and sharing). Yes, it was a relationship, when I exchanged e-mail with a stranger, and it was nice to think we were "friends" in a transcendental sense. But as time goes by these casual friendships one has with people on-line, many of which have the lifespan of a mosquito, merge with other forms of information in the bitstream, and become more like "transactions" than "relationships."

I imagined my Web site as a place where these transactions could be consummated. And if down the road it did become part of the culture to pay to visit Web sites—well, why not charge people to visit mine? I could plant a crop of "recognition," harvest it, freeze it, and sell it in special boil-in-bag pouches, just like my family's frozen creamed spinach!

At first my plan was to do the programming on my Web site myself. I bought a book called *Teach Yourself Web Publishing with HTML in a Week,* by Laura Lemay. But the time I might have invested in studying the book I instead spent on-line, and although the book remained within reach on my desk for months, it eventually found its own home near the various do-it-yourself books I bought back when we were renovating our loft, and used only to build the bookshelf they were sitting on.

So I decided to find someone to help me build my web site for me—an interior designer of virtual space. Web designers—or, as they were called when they also maintained the sites, "Webmasters" —were getting to be much in demand, especially in Manhattan. They tended to be young cyberslackers who learned HTML in college and were now finding that harried business executives, having been handed the job of setting up their companies' Web sites, were desperately trying to hire people in the know to build their Web sites for them. At the same time, a growing number of people from the print and TV media were getting into Web work. One of these people was a former book editor named Dan Levy whom I had met years before, and had recently become reacquainted with on the WELL, where he was <danlevy>, and whom I visualized as Gentle Dan, a friendly, bearlike presence. Dan lived over in the East Village, although that seemed farther away than the WELL. I sent him an e-mail explaining my problem, and asked if he would help me build my site. Although he was busy preparing a proposal to build a Web site for a division of Sony Electronics (the proposal itself was a Web site), he e-mailed me back, and offered to come over one morning and help me raise my virtual roof beam.

The night before Dan's arrival, I was sitting in my study trying to figure out how to design my site. I looked out of my real window, an operation that required no special software knowledge of any kind: I simply had to raise my eyes from my screen and look out. During the daytime the sunlight interfered with the screen, so I often kept the blinds closed, but now it was dark and the blinds were open. Across the street were two generic New York City ailanthus trees, one of which had a plastic bag caught in its branches, a particularly durable bag that would take months to blow into tatters and pennants. Next to the trees was a large metal goose-necked light-sensitive lamp, its orange glow illuminating the street. To the right of the pole was a twenty-four-hour guard in a guard booth.

The guy in the office right across the street from my window was still at work, staring into his computer screen, as he almost always was. He was a fairly new occupant of the office. He was probably in cyberspace, too, trying to capture tiny fluctuations in values of currencies across world markets, to further enrich the formidable financial combine he worked for. No friendly waves between fellow information workers had been exchanged between us, which was okay by me. Day after day we sat on either side of the river of bandwidth that ran under the street, looking at our screens, he drinking from a garden hose, me sipping through a Crazy Straw.

I pushed my padded faux corduroy chair away from my desk (by placing my two hands on the edge of my desk and pushing), and I looked around my room. I thought, This *is* my Web site. This room. Whatever I'm looking for to put on my home page is probably right around here. Floppies, pencils, paper, remote controls, the mouse, and CD jewel boxes were piled messily around the glowing screen.

My Discman. Marty the Seltzer Man's old-fashioned seltzer, in a blue bottle. A bottle of Bass Ale, empty. *PAW (Princeton Alumni Weekly)*. Books. *The Journals of Francis Parkman*—a beautiful example of the bookmaker's art. The Library of America's edition of the works of Thoreau, on acid-free paper that was supposed to last five hundred years or so. I could not imagine my PowerBook and floppy disks lasting that long. *Guide to Adirondack Trails: Eastern Region. Rosemary's Baby* by Ira Levin. *The Way of Christ,* by Jacob Boehme, the mystical shoemaker, an uneducated man whose vision of Christ was greater than any amount of reason by learned divines could achieve, and whose first great revelation occurred when he saw his reflection in a pewter dish and realized he was looking at his soul.

I had thought in going into cyberspace I was leaving the world of books behind, but as it turned out my books had been invaluable guides, and I had reread certain of them in light of my on-line experiences with the sort of urgency with which one might read the car manual when one is stopped with a flat by the side of the road. I had gone West with Francis Parkman, and seen the Oregon Trail, felt the freedom in those wide-open spaces, and then I had come East with Thoreau, and looked inside myself.

I looked at my screen. My ZTerm window was up on it, filled up with posts from the Books Conference, where I was trying to learn how to be a host. I looked at the little crawly sentences in black-and-white on my passive-matrix, three-years-out-of-date screen, and I thought, They're not "friends" in some transcendental sense, and they're not metaphors for my parents or for my experiences in boarding school, and they're not interesting artifacts of the Sixties, and they're not transactions. They're just people. It's just real life, after all. They're not some great collective brain trust, gopod, groupmind, Mind, whatever. It's neither "good" nor "bad." As Boehme says in his book, there are good things in the world, but the good things are infected with a poison. The poison is a horror, an anxiety, a melancholy, a sadness. But the poison is necessary because it gives motion to everything. The world is like a heavenly flower springing up in the universe, but the poison is always beneath its feet.

And there on the screen was one of my own postings. That was me. In theory <seabrook> could have been whoever I wanted him to be in cyberspace, but in reality he had turned out just like me. He had gone through the same conflicts with authority, the same struggle with liberal impulses and authoritarian impulses, the same conflicting influences from Western ancestors and Eastern ancestors— all of it rewound to the beginning and played all over again at twelve and a half times normal speed like one of those much-speeded-up documentary films with titles like "The Miracle of Life" in which they have the camera inserted inside the womb.

It is a strange moment when you unexpectedly see yourself in a reflective surface and don't know who you are looking at, and you are allowed a rare moment of recognition before your idea of yourself returns and the curtain closes again. It seemed amazing to me that my on-line life had followed the same course as my real life, but in the moments following that instant of apprehension I was already thinking, Of course it has to be this way, and within a few more seconds the fact that I had been amazed was the only amazing thing.

IV. A Shadow on the Floor ◊ ◊ ◊ ◊ ◊ ◊ ◊ ◊ ◊ ◊ ◊ ◊ ◊ ◊ ◊ ◊ ◊ ◊

The next day was warm and Gentle Dan arrived wearing shorts and a T-shirt and carrying his software tools and his PowerBook in a cloth bag. I showed him into my multimedia cockpit. He sat down with my PowerBook and began loading floppies onto the disk drive and installing the software necessary for me to convert to HTML the things I had written. I watched his moves on the keyboard. While he was working, Dan said, "This Web stuff is just exploding. It's kind of ridiculous. The Spin Doctors Web site I did last year would be worth over ten thousand dollars now." Dan had to speak loudly because a garbage truck was compressing trash out in the street. He went on, "I think the situation is that some people in the big-media world just view it as a problem now—like, 'We need a Web site, here's some money, go and build us one.' Yours shouldn't cost much. You ought to be able to find some neighborhood kid to do it for you, the way you'd find one to cut your lawn."

I explained my ideas for the site. The front porch, where you entered the site, would have two somewhat goofy pictures of me that were taken in a photo booth in the old Central Park zoo for my city tennis permit. From there you could go in three directions: West to the Francis Parkman room, East to the Henry David Thoreau room, or to my room. In this last room you would see a drawing I had made of the room I am sitting in now, in which so many of my travels had taken place:

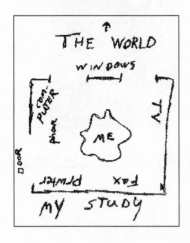

Then I showed Dan some other stuff I'd written and gave him a couple of other pictures I wanted scanned in, and discussed with him my ideas for links, all of which sounded good to him.

When he was finished converting my stuff into HTML, he asked, "Do you have a ruler?" A ruler! I got one out of a drawer in which artifacts of my prescreen work life had accumulated—a personalized letter embosser; a rubber stamp that Lisa

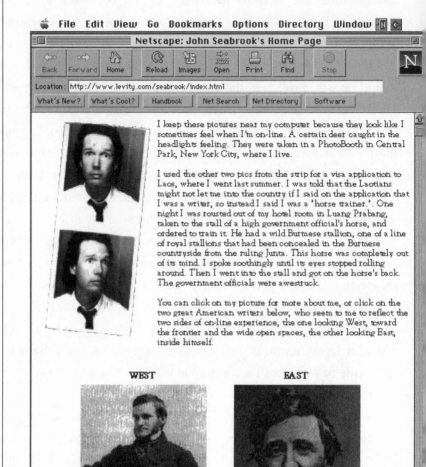

I keep these pictures near my computer because they look like I sometimes feel when I'm on-line. A certain deer caught in the headlights feeling. They were taken in a PhotoBooth in Central Park, New York City, where I live.

I used the other two pics from the strip for a visa application to Laos, where I went last summer. I was told that the Laotians might not let me into the country if I said on the application that I was a writer, so instead I said I was a 'horse trainer.'. One night I was rousted out of my hotel room in Luang Prabang, taken to the stall of a high government official's horse, and ordered to train it. He had a wild Burmese stallion, one of a line of royal stallions that had been concealed in the Burmese countryside from the ruling Junta. This horse was completely out of its mind. I spoke soothingly until its eyes stopped rolling around. Then I went into the stall and got on the horse's back. The government officials were awestruck.

You can click on my picture for more about me, or click on the two great American writers below, who seem to me to reflect the two sides of on-line experience, the one looking West, toward the frontier and the wide open spaces, the other looking East, inside himself.

WEST   **EAST**

Francis Parkman   H.D. Thoreau

carved for me; a gold Cross pen set that my dad gave me—and handed it to Dan. He laid it on my photographs, measured them, pulled the calculator down from the Apple menu on my PowerBook, and said, "Let's see. That's one and three-quarters inches, and seventy-two pixels per inch—that's a hundred and twenty-six pixels."

While Dan was working, Lisa came out of her study, which was next to mine. I watched the shadow of her door as it swung open, making changing shapes on the hallway floor. When we were designing our home Lisa and I were also getting married, and imagining the space became a way of configuring what we thought our future together would be like. What started out as an open floor plan soon grew walls, making blocks of space which the architect (who worked in pencil, on paper) called Living Area, Bedroom, Lisa's Room, John's Room. The swelling and shrinking and juxtaposition of our private spaces was the most intense part of the design process. They ended up next to each other.

And as I watched this shadow on the floor, I realized that nothing on my screen was as substantial as this shadow, or the simple arc of space that the edge of the door described. I had the feeling that I was at the end of something. The odyssey was over.

Gentle Dan uploaded my home page onto his server, tweaked my links for me, and told me my Web address—www.levity.com/ seabrook/.

"All right," Dan said. "You have a home. All within the comfort of your own home."

Chapter Eight **The Road Behind**

What demon possessed me that I behaved so well?

H. D. Thoreau

I. **Gates at the Temple** ◊

"Mind the wires, mind the wires," an usher was saying to two elderly people who had come to a synagogue on the Upper West Side on a cold night in the fall of 1995, to hear Bill Gates speak. The usher pointed to three black cables snaking in from a generator on West Eighty-third Street. The temple, apparently, was not wired for multimedia presentations, but the crowd was. It contained Gates fans of all ages, from silver-haired matrons to three teenage boys in front of me, who were talking very rapidly—kind of word-surfing off one another—about the relative merits of different Internet service providers. Gates's laptop computer waited for him on the podium, its screen glowing, and flyers that had been distributed along the benches announced the evening's theme: "Special Event—William (Bill) Gates—'The Technological Future as I See It.' "

This was also the theme of Gates's new book, *The Road Ahead,* and some people coming into the temple were carrying copies of it.

The Road Ahead sounded like the title of a campaign biography, and in a sense that's what it was. Although Gates was not running for public office, he was belatedly seeking a leadership role in cyberspace, which he called "the highway" in the book.

Audience members who'd already read the book knew what to expect from his talk: to Gates, the future looked very bright indeed. Information technology that he and others were supplying was going to make the world safer, more efficient, and full of really neat stuff, like the "wallet PC," which Gates, in a metaphor derived from his days as a Boy Scout, likened to a digital Swiss Army knife. Above all, the world would be "smart." Everything in your environment would be responsive to you and your needs. Gates's still unfinished house was an apt symbol of his vision of the future. The house sounded like an enormous computer. In his book Gates described how visitors would be issued small electronic monitors to pin to their clothing, which would communicate with the brains of the house, and he went on:

> When it's dark outside, the pin will cause a moving zone of light to accompany you through the house. Unoccupied rooms will be unlit. As you walk down a hallway, you might not notice the lights ahead of you gradually coming up to full brightness and the lights behind you fading. Music will move with you, too. It will seem to be every-where, although, in fact, other people in the house will be hearing entirely different music or nothing at all. A movie or the news will be able to follow you around the house, too. If you get a phone call, only the handset nearest you will ring.

What was interesting about this idea for living was the image of the individual bathed in his own cone of light, absorbed in his personalized soundtrack, swaddled in technology: it expressed Gates's dream of an entirely personal relationship to computers. But if my experience was any indication, the new networked world was not going to be a world in which you got to listen to your own personalized soundtrack. It seemed more likely to be a world in which other people forced you to listen to *their* music. Gates's achievement had been to make the world better conform to the needs of a person working alone, to bring more of the world within the swipe of a mouse, as it were, rather than to create spaces in which people could coexist. He was not, at least instinctively, a "network" man. Connectivity was more about groups of people working together than it was about one person giving the machine orders and having it obey. How quaint this Jetsonesque period of enthusiasm for gadgetry would appear when the real future arrived; how interestingly retro, how Fifties, this mid-Nineties infatuation with machines. The leaders of the next phase of the revolution might not even come from the computer industry. They were likely to be people who understood that this new world was fundamentally not about technology. It was about what lay beyond technology.

By eight o'clock, the temple was jammed. Helaine Geismar Katz, of the 92nd Street Y, which sponsored the event, took the podium to introduce the chairman of Microsoft, and reminded the audience that the reason we were here tonight—the main reason people were interested in Gates's vision of the technological future—was that Gates had been so right about our technological past and present. He understood how popular the PC would be, at a time when the leaders of the computer industry completely missed it, and he had the good fortune to leverage that basic insight into world power. Now that we were on the verge of another computer revolution, we were turning to a man with a track record as a visionary to tell us what it would look like.

The flaw in this logic—the assumption that the people who were the visionaries in one computer revolution were going to be the visionaries in the next one—might have been lost on some of the people in the audience, but it certainly wasn't lost on Gates himself. He began his talk by mentioning fallen giants of the computer industry; he had seen his predecessors slaughtered in front of him, and, in fact, had shared in the kill. He knew better than anyone that at the moment of a computer visionary's greatest success, the seeds of his destruction were quietly being sown somewhere else. Central shifts of insight about the nature of the computer, which in retrospect seemed so obvious—like Gates's epiphany in 1976 that everyone would want a personal computer on his or her desk—almost always went unnoticed by the leaders of the industry at the time. Ken Olsen, the visionary head of Digital Equipment Corporation, which was the Microsoft of minicomputers, said in 1977, "There is no reason anyone would want a computer in their home." Just as IBM had made the mistake of allowing the young Bill Gates to retain the rights to MS-DOS, the operating system he adapted for the IBM PC, on the shortsighted assumption that the value would be in the computers themselves, so now Gates stood in danger of reaching the limits of *his* vision, as the information revolution went through another spiral. Would he remain wedded to the operating system as his core business—Windows 95 had come out that summer, with much fanfare—while its value was eroded by the browser, and while the desktop metaphor shifted to the hypertext metaphor, in which work was not filed in stand-alone folders but linked across networks to other pages?

Gates spoke from notes on the laptop, the light of the screen blanching his face and reflecting in his glasses. Occasionally he touched a button on his machine, and a large projection screen beside him would display graphic evidence of progress in the form of the tremendous forward march of computing power, thanks to Moore's Law. I was thinking to myself that the graph did not include Lisa's computer, which she had destroyed by snagging the cord with the Christmas tree, when she was trying to drag the dried-up husk out by herself, and the machine had fallen three feet, from the dining room table to the floor. Not a long fall, but long enough. It was like Christmas in reverse—instead of finding a nice new computer under the tree, the tree ate her computer. Screen, hard drive, everything ruined, and all information that wasn't backed up (our financial records, her writing) was lost. Twenty-five hundred bucks down the drain. The guy at Tekserve, on Twenty-third Street, told her the Mac 180's sorta were like that . . . drop 'em and they die.

Gates could probably have said something extremely interesting in his talk about how the trend toward networked computing will affect our view of privacy on-line, or about the social revolution that might occur because of the much cheaper access to computing power, or about all the political problems you have when a bunch of isolated cones of light meet. But instead, as in his book, he settled for the august tones of the professional futurist and a weary-sounding rhetoric of optimism. "The Industrial Revolution was nothing compared to what we are talking about here," he said, and "We don't call it a mania—we call it a gold rush." Technical progress would make people wealthier, happier, healthier, more sociable, more able to do what they want. It would solve traffic problems in crowded places and would make people who were alone and had nowhere to go feel connected. As for concerns about security and information overload: "Those things are being worked on." His faith in technological progress was indeed like a religion, though not a religion for the technologically lame or the un-smart.

I had been inclined to believe Gates two years ago, when he expressed similar ideas in his e-mail to me, but having spent the intervening years in the belly of the progress machine, I found myself doubting him tonight. Yes, progress of a purely technical nature is inevitable, and if everyone confined themselves to making pronouncements on technical progress alone, there would be no problem. But people feel compelled, for all kinds of different reasons, to infer social progress from technical progress—to believe that better machines will make better people. This sounds absurd when stated baldly, which is why it is usually only implied indirectly, by way of metaphor. The assumption at the heart of the many-to-many metaphor, for example, was that a system designed to promote free speech among machines will promote free speech among people too. In my experience this turned out to be only half true. But it *sounds* logical. Why should it not be the case, when the age, sex, and race of one's correspondents do not come through the wires unless they choose to reveal such information, that the medium will help rid the world of ageism, sexism, and racism? But in fact it turns out that these and other prejudices are so deeply ingrained in humanity that even when egalitarianism is forced on users by the technical limitations of a medium, people find a way to be just as cliquish and exclusionary as they ever were.

The closest Gates came to discussing social issues was when he told a favorite anecdote. "Last week, I went out to dinner," he said. "Someone came up to me, asking me for money. I wasn't sure what to do. Then he said that I should look up his home page, and gave me his URL. I didn't know if it was true or not, but it was such a good line I gave the guy five bucks. So he may be a homeless person with a home page."

I glimpsed <billg> only once, at the end of the evening, when Gates read audience questions from slips of paper that had been passed up to him. One of them said, "Which of your competitors have the best strategies for the Internet?" Gates gave a pained laugh, and then responded in <billg> mode: "That's sort of like, Tell me where I'm going to die, and I'll be sure not to go there."

II. go books ◊

As a co-host of the Books Conference on the WELL, I try to create a stable platform on which chance meetings, coincidences, and collaborations can take place. When I see newbies wandering around in the kind of whiteout I remember from my newbie days, I try to point them toward the trail. My beloved co-host, <rbr>, actually does a lot more of the grunt work of hosting than I do, like changing "the mystery log-in quote" and linking interesting book-related topics from other conferences to ours, but hopefully I'll have more time to devote to the conference in the future.

I keep an eye on the conference hits. Every time someone who is a member of the WELL types "go books," the groupmind scores one hit for us. Hit counts are tabulated daily, and monthly rankings of our conference in relation to the other conferences are available by typing "stats" when you're in Backstage—the private conference for Hosts. When I started co-hosting, our conference dropped from seventh rank down to ninth, which made me feel pretty bad, but since then we have climbed up to fourth. The success of our conference gives me a low buzz of "recognition" that helps get me through the day. Our ascendance came partly on the strength of a beautiful thrash we had awhile back, started by the appropriately named <woodrow>, whose real name was Abe Opincar, which began with this posting:

Topic 888: I Almost Shit a Brick: Anne Fucking Rice on Kafka
Started by: Oh, No, Not Another Fucking Elf (woodrow)

In her hot little hands a friend brings to me tonight *two* new paperback Schocken editions of Kafka. One of them, *The Metamorphosis, In the Penal Colony, and Other Stories* includes a foreword by ANNE FUCKING RICE!!! That's right, ANNE FUCK-

ING RICE. She says his are "among the finest horror stories ever written."

Call me an elitist swine, but I'll be goddamned before I sit quiet and let the likes of ANNE FUCKING RICE pontificate on St. Franz.

Aux armes, enfants, aux armes! As far as I'm concerned, this is the first shot heard 'round the world in the fucking culture wars.

A line has been crossed. This means war.

C'est la fucking guerre!

From that posting ensued an excellent fucking thrash as far as this Host was concerned, involving wide swings among the posters, with some new posters pitching in, though for the most part the posters were the usual suspects. In theory, I as a host have the authority to stop thrashes, and would try to stop one if the thrash showed signs of getting "out of hand," meaning if a pern's feelings might be getting too badly hurt. But in fact hosts rarely stop thrashes, and some hosts actually encourage them, because they are so good for the hits. And positive things sometimes come from these on-line dramas. In this case, another user, <cthroid>, who is in real life the writer Cynthia Heimel, noticed <woodrow>'s posting, and e-mailed him about it; they met IRL, and by the following summer they were married to each other. Our first Books marriage, in my time as a host! I'm so happy about that.

I have learned to back away from on-line fights, and to make overstated displays of submissiveness in front of the would-be alpha pern, and I have even started using smileys—:)—and grin signs— <g>—in my postings, which I never thought I would do. I try to talk about what's going on inside the thread. It's a relatively easy way of adding "content" to the thread but at the same time of staying out of trouble. This is perhaps a little bleak, given the utopian fantasies I once cherished for this medium, but I have spent too many hours in bitter argument with angry antagonists, when in a face-to-face meeting we would have resolved our differences in five minutes, to believe in the utopian promises of this technology anymore. Whatever capacity the medium has for bringing people together, it has an equal capacity for driving them apart; and the solace one may find on-line is offset by malice, and the compassion by cruelty, and the goodwill by spite. It appears that, at least in this instance, God does appear to make the world restless, to move and revolve in all its parts, only to come to the same place again, just as Jonathan Edwards feared.

One night not long ago, as Lisa and I were walking along a dark Tribeca street to a local hangout, I was telling her about the <woodrow>-inspired thrash, and saying that it was a "four-lawnchair," thrash for the lurkers, according to a system devised by <booter> to measure lurker interest, when Lisa said,

"So, in other words, all this stuff about the Net being a utopia is over now."

Lisa, as it happened, was in the early stages of her own on-line adventures. She had recently made her first posting—a ringing defense of President Aristide—in the Usenet newsgroup soc.culture.-haiti, and now I frequently found myself alone in the living room, reading a book, while Lisa sat in front of her screen reading the latest postings to the newsgroup, and exclaiming, "Hoo hoo hoo!" and "Gloryoski!" Some fellow supporter of Aristide's whom she had met in the newsgroup was now sending her e-mail suggesting they make "common cause" together. Back off, buddy.

Anyway, I didn't want to rain on her parade, so I said,

"No, not entirely, but the fact is that a little provocative commentary really gets the hit count up."

"So basically, what you're saying is you're just like some schlock TV producer now. Like 'A Current Affair' or something."

"More like Sally Jessy Raphaël, actually."

"Whatever."

The Owner's efforts to grow the WELL continue apace. We now have a multimedia interface called "WELL Engaged" (gag me with a spoon), which lets you read some conferences on the WELL from the Web, and is clickable, and has color and graphics to supplement the ASCII text. The new GUI was of course the source of a thrash in the Policy Conference, with the usual accusations about how the management doesn't give a shit about the community—though now the "community" was referred to as "the content providers" by some posters, which I took as a sign of the times. <katz> got a fair roasting in that topic but failed to lose his temper and tell anyone to fuck off, and the thrash died down after three hundred or so postings. As far as I am concerned, WELL Engaged is slower and clumsier and less useful than Picospan, and, as a word person who is really interested only in the conversation, I couldn't care less about the multimedia aspects, but this is Progress, and supposedly it will bring lots of new people onto the WELL, so what the heck.

Sometimes I feel like the character at the end of Evelyn Waugh's novel *A Handful of Dust,* condemned to read Dickens for the rest of his life to a blind man in the middle of the Brazilian jungle. I have fantasies of logging off forever and of bricking <seabrook> up in a wall in the tower stairwell (lately, for example, I have caught him engaging in some of the same bogus hostly behavior for which I used to mock my father at the dinner table—it's like <seabrook>'s grown *older* than me now). But I never actually do leave, and I doubt that I ever will. I always return from these fugue states with a new resolve to bury my flame ghouls decently, put as much time as I can spare into hosting, and help other people if I can—if I read a good book, post about it; or if I hear of some work to be done and know someone else who needs work, try to put the two together; or try to be gracious when some new on-line "friend" sends me e-mail. It is a pleasure to participate in a system you help to run. I'm sure that I have many more levels of cyberunderstanding left to achieve, and that some helpful pern will point that out in the Books Conference. (At least it will be good for the hits.) I pray for the patience to accept their criticism, and for the grace to suspend myself in the middle of the thread, like a trout hovering in the middle of a bitstream, with almost invisible fin and tail flicks, wise enough to tell the food from the tied flies.

III. **In the Palace** ◊

In the end, it doesn't really matter whether the on-line world represents progress or not. The giant electronic brain—half human, half machine—grows larger. Even Bill Gates is its puppet. It has no moral purpose that I can discern, but, like me, most people will not be able to refuse it, so you may as well advance confidently and hopefully in the only direction that makes any sense—deeper. It's not progress, but it's movement of some kind. Maybe you will meet with success unexpected in common hours, or maybe you will find yourself in a high-tech airplane that has just been blown in two by a high-tech bomb, afforded twenty-four seconds of terror before hurtling into the kind of impact that will instantly make you a foot shorter.

I still get pretty excited about the latest thing to hit the Net. Not long ago, while working at the *New Yorker,* I downloaded a software "client" called "The Palace" and installed it on my desktop machine. The Palace was a kind of virtual reality software for Chat, developed by Time Warner, which allowed you to put your own image, or what the digital guys were calling your "avatar," into a 3-D environment, where you could see the avatars of other people in the chat room. It was an early step toward establishing spatial relationships with other people on-line, which would be a big step forward. After waiting with the usual intense irritation for almost a half hour while the "client" software downloaded, and then spending another fifteen minutes trying to get to the Palace Web site, I had finally made it into the Palace, and was feeling again the old lightness of hope and possibility—maybe *this* is the place—when just then the editor of a piece I had written came in to "close" it: that is, to make the final round of corrections and changes, before shipping it off to the printer, in Kentucky. Having spent so long in the Palace queue, I didn't want to lose my place, so I left the Palace running on the screen while we got to work on the galleys.

A "closing" at the *New Yorker*—generally conducted by the writer, the editor, and a representative of the copy department whose enormous authority was cleverly disguised in the vague-sounding title of "OK'er"—was a beautiful act of print culture, my culture, for which I knew no parallel on-line. The Net was a wonderful medium for the creation of new words, but there was nothing quite like this, the preservation of a certain way of using words, done under the pressure of "real time"—linked to production schedules, publication dates, quittin' time, and other aspects of the print world that on-line hasn't been around long enough to develop, and maybe never will. We all worked in pencil, that marvelous tool, our heads bent over big, sumptuous sheets of paper, silently reading the columns of type. I admired the old, arcane symbols of the typesetter's art, remembering that this was how I met my wife—she copyedited one of my articles. How familiar, how pleasant this was, the pleasure of choosing the right word, of taking the time to hunt for it before setting it forever in print, stressing yourself in concentration—a stress that is a uniquely taxing form of exercise.

But suddenly the blissful decency of the moment was shattered by the sound of a fart! The editor and OK'er were both jerked from their mild and patient regard of my piece, stricken looks on their faces. The offensive sound seemed to have come from behind my desk, which happened to be where I had my slender Quadra 610 sitting (between the end of my desk and the wall) but which was also, of course, the area that a real fart would come from, had I been indecent enough to let one rip.

"It was the computer!" I said, pointy-point-pointing at my screen, like you would point at the dog when the dog actually *did* do it. Among the innovations in the Palace virtual reality software, apparently, was the ability to hear other avatars fart in the public space that your avatar occupied. Progress! Like the good netizen I have become, I instinctively thought "Software controls" and, taking hold of the mouse, dragged the cursor up to the top of the screen, clicked on one of the menus, and dragged down to "Preferences."

"Let's see, there must be some way to turn the farter off. . . ."

IV. The Spearfish Valley ◊ ◊ ◊ ◊ ◊ ◊ ◊ ◊ ◊ ◊ ◊ ◊ ◊ ◊ ◊ ◊ ◊ ◊

When Lisa and I got married we spent the first two nights of our honeymoon in a bedroom on a train, traveling west from Albany to Chicago on the first night, then northwest across the northern plains on the second. Both our mothers' grandparents—in her case, the Fitzpatricks; in mine, the Toomeys—had come out West and homesteaded places in the Dakotas in the 1870s, and we had been wondering if our ancestors' paths might have crossed somewhere out there in the great American dreamland, because like cyberspace the West was a small place in those days. In the nighttime we sat side-by-side in our little conjugal box, like two monkeys in a Sputnik, watching as the black and haunted-looking land unspooled outside of our window. All but the brightest traces of civilization had melted away, and the immense size and mystery of the land loomed up before us. Destiny! Our destiny! It was somewhere out there in the darkness, waiting for us to discover it. . . .

During my two years of sitting here in front of my screen, I have sometimes wondered whether destiny has, in its joshing-around way, brought me back to within a few miles of the street in Brooklyn where my great-grandfather started out on his grand adventure, plunked me down here in front of my screen, and somehow compelled me to repeat his experience, in metaphor, in cyberspace. It was as though my ancestor's story was a meme that had propagated itself in me. Just as D. J. Toomey probably did, I had heard colorful accounts of the frontier, depicting it to be a place of action and charm and vast opportunity, and I was restless and eager to better my own circumstances, and if possible to build up the kingdom of God, too, so I also left home. But in my case I only had to ride my bike down to the local computer store and buy a modem, then come back here and figure out how to make it work. I didn't need physical courage to get around the frontier, and this strikes me as sort of sad. To find the frontier these days, you have to look inside your own mind, and while that is interesting, and certainly more convenient, it isn't the same thing.

In his fifth year of tramping around the West, D.J. went to Deadwood, South Dakota, where he had heard there was gold. On the way he got lost and walked into the camp of Spotted Tail, the great Sioux warrior, but escaped without anyone getting killed. In Deadwood he found no gold, but he did see Wyatt Earp, Wild Bill Hickok, and Calamity Jane, of whose boisterous behavior he disapproved. He also heard from an old prospector named Frank Bryant of a place in the Black Hills called the Spearfish Valley, "the prettiest spot on earth," the prospector said, "equal, if not superior to the garden of Eden." That sounded like the place he was looking for, so he and a partner went to see it.

> On May 1, 1876, a Mr. Gus Smelzer and I saddled our ponies and packed some grub and tools at the mouth of Blacktail, went down False Bottom to the open prairie, thence west to Spearfish. We had neither road nor trail to guide us, only direction and instinct. Our first view of the valley was from what is now cemetery hill and it was surely a pretty picture as it looked on that day in May, in its natural and virgin beauty. No sign of settlement, no trees, only a light fringe of oaks, ash and elm, with a few cottonwoods bordering the stream. A fire the previous fall had cleared the old grass. The new was showing green as far as the eye could see. I have often wished I could paint a picture of that scene as I saw it then and still see it with my mind's eye. I right there and then agreed with Frank Bryant that it was a gem and my future home for life.

Like me, D.J. was often called on to talk about the rogue hackers and flamers he must have faced in his time on the frontier. "I presume you will expect me to tell some hair-raising stories of the Indian trouble," he says in another *Queen City Mail* story, "but as a matter of fact we were never in any great danger from the Indians, provided, of course, that we kept prepared and were careful to avoid being ambushed." He was amused by his public's need for these heroic tales. "You will hear expressions of sympathy for the hardships the early pioneers suffered," he says. "That is mostly bunk." He did describe in fine detail a paranoid episode that his partner Gus Smelzer had during their first visit to the Spearfish Valley:

> We had been camping and working together six or seven days. He seemed normal until the last night. We made our bed as usual in the willows away from our camp fire, a precaution for protecting us from Indians. I noticed him acting queer while cooking and eating supper.

I wanted to stay a day or two longer and put a few more logs on my cabin. It was only four logs high while his was eight or ten high. So he finally consented to stay one more day, although he said we were taking desperate chances. He was not naturally a coward or he would not have come out here alone, which he had done a few days previous to my meeting him.

He went to bed with all his clothes and boots on, tied his horse close up with saddle on. Shortly after lying down he commenced hearing things, a mouse in the dead leaves, a rabbit hopping, the brush seemed to him like Indians closing in on us, and finally to keep him from bolting I agreed to pack up and pull out. It was perhaps midnight when we crossed the creek and went up along the Spring Creek toward the Hills. It was a pitch dark night. We walked, leading our horses packed with our bedding, tools and camp plunder, and a deer or two. We hadn't gone a mile when he began looking back and could see the Indians following us. I had a hard struggle to keep him from dumping things, jumping his horse and running off. A clear case of shattered nerves, caused, I believe now, by too much strong coffee, although at the time I did not know why, but knew I had a part lunatic on my hands. That experience convinced me that the condition of one's nerves makes cowards or braves of all of us.

Likewise, I was neither hacked nor invaded by a virus, although I did have a paranoid episode in which I thought my computer was infected with a virus, and I had lots of chances to reflect on the truth of the remark that the condition of one's nerves makes cowards or braves of all of us.

The Black Hills were the very last part of the Western frontier, the end of the line for the pioneers, and D.J. got there just in time. When he posted his claim in the Spearfish Valley, in May 1876, he did not see a single white man or settlement in the area. When he returned four months later to prove up his claim, he saw a log cabin built every half mile, including another man's cabin on the site of his 160-acre homestead. The Web had arrived. "People were coming in by the thousands every day. Deadwood and Gayville, the only towns started yet, were growing fast. The placer mines in Deadwood Gulch, the first to start sluicing, were producing well." He found the men who had built the cabin on his claim, and told them that if it wasn't removed in ten days he would set fire to it. "I could see that that they were impressed." The structure was removed, and D.J. built himself another cabin on the place.

Same here, D.J. I got on-line before hardly anyone heard of the Web, and by the end the Net *was* the Web. Maybe the word "Net" would be preserved for aspects of the on-line world more in keeping with the true spirit of the many-to-many, but Web is the word everyone wants to use now.

D.J. married the first schoolteacher to come into that country, Vesta Wales Noyes, and they had six children. One day, having come home from hunting, he was sitting in a chair, using the barrel of his gun as an arm rest, when his baby daughter Maude toddled over and pulled the trigger, and the blast blew off his arm above the elbow. Even with one arm, however, he couldn't sit still. In 1889 he heard about the gold strike in the Klondike, and, leaving Vesta to look after the ranch and the children, took off with a partner, two wagons, five horses, and twelve hundred pounds of freight, going overland to Edmonton, Alberta, and from there west to the MacKenzie River, thinking that they could then make it over the Chilkoot Pass and down into Dawson City. But they lost the horses somewhere west of Edmonton and split up when they reached the head of the Finley River. D.J. built a canoe and came down the Finley and the Peace rivers by himself, paddling and portaging six hundred miles with one arm. He was gone for two years and returned with no money, nothing to show for his pains but a wonderful set of chess pieces he had whittled himself during his two years in the wilderness, and which my mother now keeps inside an old Bonwit's box, on the shelves behind the door in the Silk Room. According to a note from my mother that is in the box, D.J. managed the task by holding down a piece with his stump and whittling with his good arm. What little gold he had found he had made into two rings, which he presented to his two daughters. But Vesta was furious with him for leaving her to cope with the ranch, and the following year they sold the homestead and moved to a house in Spearfish town, where D.J. went into the dry goods business, and where my mother, his grandchild, was born, in 1924. When he retired, D.J. and Vesta moved to San Diego, where he died in 1942, at the age of ninety, a year after Vesta died. "From the time I was a small boy," he said,

> I wanted to live on a farm with a running stream and lots of trees and bushes for birds and rabbits, but I never found the place until May 1, 1876. That was the day I first laid eyes on our old ranch in Spearfish Valley. When I drove the first claim stake locating 160 acres of land I said then that nobody but me would ever own that place as long as I lived. I've been sorry lots of times I did not stick to that promise.

One night not long ago I got the urge to talk to my mother about him. It was far too late to talk on the phone, so I wrote her an e-mail. How convenient and pleasant it was, just to log on and talk to Mom in the asynchronous eternity of e-mail.

> Hi Mom, I was just reading over some of those old stories of your grandfather's that you sent me a while ago. They're pretty great. What a life he had. But it was sad, the way it ended, with his losing the ranch. He seems to have had wanderlust. Why else, after he was settled at the ranch, would he have taken off for the Klondike like that? What was he thinking? What was he looking for?

Mom, who had recently returned from visiting the Spearfish Valley, which she found full of fast-food joints, cheap motels, malls, and other signs of progress, responded with her usual timeliness.

From: <72733.3174@compuserve.com>

Maybe it was a midlife crisis that made him go to the Klondike. I guess he had several children by then, or maybe all of them, and felt overwhelmed. I honestly don't know if looking for gold or looking for adventure were his main purposes in heading for the Yukon. A bit of both, I suspect. It was obviously regarded as a damn fool idea by his wife. I think she maybe never forgave him for that two-year absence. When I knew them, later on, they seemed to live in a state of armed truce. She "humphed" a lot when he said something.

His wife was extremely capable, and much better educated than he was. I have only one book of my grandmother's, and it is a philosophy book. She could help us with our French and our algebra when she came to visit, and did crossword puzzles all the time and knew all the Latin words which used to impress me enormously. He, on the other hand, liked to wander around town and chat with people.

Perhaps you connect somehow with the desire to find out about new places and new things. D.J. had the first radio set in Spearfish, which he got in 1919. He also, in 1914, was driving his own car, with his one hand managing to shift gears and steer and put on the hand brake. My mother described a terrifying ride from Deadwood, where the train stopped, to Spearfish when she and my father went to Spearfish from St. Louis on their honeymoon in 1914.

The eagerness to try new things seems to be a gene that has been passed down. It might be said to be a spirit of adventure and it could be described as poor judgment. My father bought an airplane and learned to fly in 1928, even though his sister, Maude, had been killed in 1918 or 1919 stunt flying with some barnstorming pilot who came through the area. Maude was an excellent school teacher, about 30 when she was killed. An early Christy McAulliffe (sp?)

I suspect D.J. would have wanted to learn about cyberspace too. He was nearly illiterate, though, having had the benefit of a few years of Catholic schooling in Brooklyn. He left both the city and the church, partly because he hated being treated so cruelly by the Catholic priest or priests. He said he got beaten by one of the priests and decided no one had the right to beat another human being.

He was probably not dumb, though. He also was respected as a man of his word.

Thanks, Mom. Glad for the information.

Loss. That was part of the map that D.J. had left behind. The road to Illumination led through the Land of Loss. D.J. had found the place, but then he had lost the place. That was the way I had gone, too—loss of innocence, loss of illusions, loss of the frontier, and trying to make up for those losses and other losses that had come before me. Sometimes I have the strange feeling that everything that happened to me in my two years on-line actually happened long before I was even born, and what I have really been looking for in cyberspace is something that somebody else lost a long time ago.

New York City
June 1996

Endnotes ◆ ◆

Just as a February expedition into the Adirondack backcountry would not be possible without skis or snowshoes, so this excursion into c-space would have been impossible without the following references.

Preface: The Autobiography of Joe Homepage

13 mountain men: Jim Bridger, Milton Sublette, and Tom Fitzpatrick, among others, were the Net dot gods of their day. Their adventures in the great American wilderness of the 1830s are memorably told by Bernard DeVoto in *Across the Wide Missouri* (New York: Houghton Mifflin, 1975), a book I carried around with me during the early part of my travels, until I put it on the roof of the car while packing to go winter camping in the White Mountains of New Hampshire, and by the time I remembered I'd left it up there I was on the Henry Hudson Parkway, and it was gone.

> The mountain men were a tough race, as many selective breeds of Americans have had to be; their courage, skill and mastery of the conditions of their chosen life were absolute or they would not have been here. Nor would they have been here if they had not responded to the loveliness of the country and found in their way of life something precious beyond safety, gain, comfort, and family life. Moreover, solitude had given them a surpassing gift of friendship and simple survival proved the sharpness of their wits. There were few books and few trappers were given to reading what there were: talk was everything. The Americans, and especially the Americans who live in the open, have always been storytellers—one need recall only the rivermen, the lumberjacks, the cowmen, or in fact the loafers round any stove at the rural crossroads [or in the newsgroups, chat rooms,

etc.]. Most of their yarning has been lost to history, but it was a chronicle of every watercourse, peak, park, and gulch in a million square miles, a chronicle of chance happening suddenly and expectation reversed, of violent action, violent danger, violent mirth. How one who was with us last year was eviscerated by a grizzly or gutshot by a Blackfoot. How we came into Taos or the Pueblo or Los Angeles and the willing women there and the brandy we drank and the horses we stole. (pp. 44–45)

Part One: West

17 "And yet stern and wild associations gave a singular interest to the view . . .": Francis Parkman, Jr., *The Oregon Trail* (New York: Penguin Classics, 1983), pp. 105–106.

This quotation follows the first (1849) edition of Parkman's book, not the author's 1872 revision, in which the last two sentences of the passage were omitted because Parkman no longer believed in the noble qualities of the wilderness. "Almost all contrasts between the effeminate East and the robust West are deleted in 72," writes E. N. Feltskog in his authoritative text of *The Oregon Trail* (Madison: University of Wisconsin Press, 1969). In the early editions of the book, writes Feltskog,

> The East—and especially New England—is a haven of peace and rest. The West is destructive; it brutalizes white men and Indians alike and reduces civilization to a mocking memory. Yet the West tests a man's endurance and proves his courage and strength; the East saps moral and physical energy by offering him safety without effort, experience without pain. (p. 50a).

In later editions, Parkman toned down the contrast between West and East, perhaps because in his old age and sickness his ideas of the West were more sensible of "the terrors of the strange and forbiddingly vacant mind of the primeval world," while his visions of the East "become at last indistinct and melancholy, the desolate memories of an irretrievably lost sanctuary where Nature was far kinder and softer than the empty plains and dark mountains he crossed with the Oglala village" (pp. 50a–51a).

Bernard DeVoto, in his book *The Year of Decision: 1846,* argues that Parkman, in going west in 1846 to admire the landscape, hunt buffalo, and observe Indians, like other young gentlemen of the day, missed one of the greatest historical events of the nineteenth century: the Great Migration.

> It was Parkman's fortune to witness and take part in one of the greatest national experiences, at the moment and site of its occurrence. It is our misfortune that he did not understand the smallest part of it. No other historian, not even Xenophon, has ever had so magnificent an opportunity: Parkman did not even know it was there. (p. 115)

Mason Wade, in his introduction to Parkman's Oregon Trail notebooks, writes, "The easy friendliness of the West grated on [Parkman's] Bostonian reticence, and he was happier with the Indians and with the French-Canadian halfbreeds, to whom he could adopt a superior attitude, than with his fellow Anglo-Americans, who would not tolerate such an attitude" (*The Journals of Francis Parkman* [New York: Harper Brothers, 1947], p. 403). The emigrants, whom Parkman calls "offscourings of the frontier," and whom he found "totally devoid of any sense of delicacy or propriety," were creating a new nation right under his nose—and he couldn't see it. Writes DeVoto: "Manifest Destiny was taking flesh, his countrymen were pulling the map into accord with the logic of geography, but they were of the wrong caste and the historian wanted to see some Indians." DeVoto "aches for the book that might have been added to our literature if God had a little thawed the Brahmin snobberies." But no: "The historian succumbed to a parochialism of his class and we lost a great book" (pp. 142, 175–176).

Chapter One: The Nerd Within

24 "every new way of getting wealth . . .": Alexis de Tocqueville, *Democracy in America* (New York: Anchor Books, 1969), p. 462.

25 "Does God make the world restless . . .": Edwards is quoted in that old chestnut of the American Studies major, Perry Miller's *Errand into the Wilderness* (Cambridge: Harvard University Press, 1956), p. 234.

27 "By turns the steamboat . . .": Lewis Mumford, *The Pentagon of Power* (New York: Harcourt Brace, 1970), p. 296.

Mumford is the granddaddy of liberal skeptics of the progressive promises of technology, and his influence can be seen in other books that, like his, first ran in the pages of the *New Yorker:* Rachel Carson on the adverse effect of petrochemicals *(Silent Spring)*, Paul Brodeur on the harmful aspects of microwave fields *(The Zapping of America)* and on electromagnetic radiation *(Currents of Death)*, Jonathan Schell on nuclear destruction *(The Fate of the Earth)*, and Bill McKibbon on the greenhouse effect *(The End of Nature)*.

27 the philosophy of progress: I love books about the history of technological innovation that are written from the evolutionary point of view. Sadly, there aren't many of them; most American historians of technology seem to prefer the "discontinuous" or "great man" theory of invention. Two exceptions are George Basalla's *The Evolution of Technology* (New York: Cambridge University Press, 1988) and William Greenleaf's *Monopoly on Wheels: Henry Ford and the Selden Automobile Patent* (Detroit: Wayne State University Press, 1961).

33 "The new age of intelligent machines . . .": George Gilder, *Microcosm* (New York: Simon and Schuster, 1989), p. 381.

There are many other similar bursts of what used to be called "the technological sublime" in Gilder's bit-soaked, feverish futurism—which has a long past in American letters. For example, compare this piece of Gilder:

> The overthrow of matter in business will reverberate through geopolitics and exalt the nations in command of creative minds over the nations in command of land and resources. Military power will accrue more and more to the masters of information technology. Finally, the overthrow of matter will stultify all materialist philosophy and open new vistas of human imagination and moral revival. (*Microcosm*, p. 18)

to the transcendentalist George Ripley, writing in 1846:

> The age that is to witness a rail road between the Atlantic and the Pacific, as a grand material type of the unity of nations, will also behold a social organization, productive of moral and spiritual results, whose sublime and beneficent character will eclipse even the glory of those colossal achievements which send messages of fire over the mountain tops, and connect ocean with ocean by iron and granite bands.

Or compare the first sentences of the Progress & Freedom Foundation's manifesto "A Magna Carta for the Knowledge Age," of which Gilder is a coauthor (with Alvin Toffler, Esther Dyson, and George Keyworth)—"The central event of the 20th century is the overthrow of matter. . . . The powers of mind are everywhere ascendant over the brute force of things"—to John Walker, writing in praise of mechanical invention in the *North American Review* in 1831: "From a ministering servant to matter, mind has become the powerful lord of matter." Meanwhile, matter continues to hang tough. . . .

36   "I realized later . . .": Bill Gates, *The Road Ahead* (New York: Viking, 1995), pp. 1–2.

36   "He was always upset . . .": Stephen Manes and Paul Andrews, *Gates: How Microsoft's Mogul Reinvented an Industry—and Made Himself the Richest Man in America* (New York: Doubleday, 1993), pp. 21 and 442.

Chapter Two: <billg>, BMOC

48   Perhaps it is true: Jacques Ellul, *The Technological Society* (New York: Vintage Books, 1964), pp. 24–25.

54   "My vision of the future . . .": *Princeton Alumni Weekly*, December 20, 1995, p. 9.

65   encounters with grizzly bears: Three books served as useful guides, not only in this encounter with the enraged Bill Gates, but in subsequent run-ins with other poorly socialized individuals on the Net: Andy Russell's *Grizzly Country* (New York: Nick Lyon Books, 1967); *The Great Bear*, a collection edited by John A. Murray (Anchorage: Alaska Northwest Books, 1992); and Stephen Herrero's

Bear Attacks: Their Causes and Avoidance (New York: Lyons & Burford, 1985). If, after you have stood your ground before a grizzer, and looked down and away, the bear decides to attack you anyway, Herrero advocates passive resistance. "Passively resisting the attack by remaining as motionless and soundless as possible—playing dead—seems to have decreased the intensity of injury," he writes. "In exceptional cases people have stunned such attacking grizzly bears by hitting them with a club, or just their fists or knees, or by sticking their fingers in a grizzly's large nostrils, and the bear has left. Because grizzlies are stronger and have superior natural weapons, it is usually the human being who takes the worst punishment in a fight with an attacking bear" (pp. 27–30).

Chapter Three: The Great Migration

86 "Great is journalism . . .": Carlyle is quoted in Elizabeth L. Eisenstein's interesting essay "The End of the Book?" in *The American Scholar,* vol. 64, no. 4, Autumn 1995. Carlyle was one of the first writers in English to question the assumptions of progress at the heart of the "technological sublime," by wondering whether "[m]en are grown mechanical in head and in heart, as well as in hand." His first great work, the antitechnology essay "Signs of the Times" (1829), speaks to our times too. Would that the Unabomber had Carlyle's eloquence. . . .

88 "The technology of these transformative systems . . .": Roy Ascott, "Connectivity: Art and Interactive Communications," in *Leonardo,* vol. 24, no. 2, 1991, p. 116.

89 "Americans of all ages . . .": Tocqueville, *Democracy in America,* p. 513.

90 Mysteries of Skill: In August 1858, Ralph Waldo Emerson was with a party of philosophizing nature lovers on Tupper Lake, in upstate New York, when someone brought the news that the first transatlantic cable had been completed. Emerson celebrated the news by writing in his journal a poem, "The Adirondacs":

> What in the desert was impossible
> Within four walls is possible again,—
> Culture and libraries, mysteries of skill,
> Traditioned fame of masters, eager strife
> Of keen competing youths, joined or alone
> To outdo each other and extort applause.
> Mind wakes a new-born giant from her sleep.
> Twirl the old wheels! Time takes fresh start again,
> On for a thousand years of genius more.

But what if Emerson had heard word of the transatlantic cable on America

Online, prefaced by a cheery electronic voice saying, "Welcome! You've got mail!"? Or what if the news had arrived on a cell phone that one of the philosophers had brought along in the canoe? Would <billg>, the Sage of Redmond (I hear America hurling), have moved the Sage of Concord to such optimism about the future of mass communication? But aren't e-mail, cell phones, and Windows 95 logical extensions of the transatlantic cable, products of the thousand years of genius that Emerson was calling for?

Chapter Four: My First Flame

115   loonier and loonier posts: I am indebted to David Delaney's wonderful collection of net dot loons, in his net.legends.faq.

Part Two: East

127   "It follows that the wise revolutionary legislator . . .": Isaiah Berlin, *Four Essays on Liberty* (New York: Oxford University Press, 1969), p. 20.

Chapter Five: I Am a Node on the Net

130   the nineteenth-century pioneer experience: "This then is the heritage of pioneer experience—a passionate belief that a democracy was possible which should leave the individual a part to play in free society and not make him a cog in a machine operated from above; which trusted the common man, in his tolerance, his ability to adjust [to] differences with good humor, and to work out an American type from the contributions of all nations—a type for which he would fight against those who challenged it in arms, and for which in time of war he would make sacrifice, even the temporary sacrifice of individual freedom and his life, lest that freedom be lost forever" (Frederick J. Turner, *The Frontier in American History* [New York: Henry Holt, 1962], p. 358).

148   How could Jefferson not have foreseen: See Leo Marx, *The Machine in the Garden* (New York: Oxford University Press, 1964), pp. 116–144.

149   "In all conversation between two persons . . .": Ralph Waldo Emerson, *Essays: First and Second Series* (New York: Library of America, 1990), p. 160.

149   "The sensation of personally participating . . .": Howard Rheingold, *The Virtual Community* (New York: Addison-Wesley, 1993), p. 110.

152   "a unit of cultural transmission . . .": Richard Dawkins, *The Selfish Gene* (New York: Oxford University Press, 1976), p. 206.

157 "a society of liberal, intelligent and cultivated persons . . .": Quoted in Annette Kolodny's introduction to Nathaniel Hawthorne's *The Blithesdale Romance* (New York: Peguin Classics, 1986), p. vii.
173 "Now roam in the depths of hell . . .": Snoop Doggy Dogg, "Serial Killa," from the album *Doggystyle*. Mr. Dogg's "murderous styles and poetical techniques" are a good way to counteract the effects of reading too much Hawthorne and Emerson.

Chapter Six: The Post and the Thread

200 "sitting in front of our computer . . .": Rheingold, *Virtual Community*, p. 19.

Chapter Eight: The Road Behind

251 "When it's dark outside . . .": Gates, *Road Ahead*, p. 218.
262 the first two nights of our honeymoon: When I wasn't doing it in the style advocated by the sexy girl group TLC in their song "This Is How It Works," I was reading Matthew Josephson's *The Robber Barons* (New York: Harcourt Brace, 1962), another great work of narrative American history.

Josephson was associated with the New York surrealists in the 1930s, a set that included William Seabrook, the author of *The Magic Kingdom*, a crazed book about voodoo in Haiti, who is supposedly a distant relative of mine. "Willy Seabrook? Oh yes, I knew him well," the artist Al Hirschfeld, who also hung with that posse, once said to me when I was seated next to him at a black-tie event in Sardi's restaurant, while I was reporting a story about Cindy Adams, the *New York Post* columnist. "He was a cannibal, you know," Hirschfeld continued.

"Is that right?"

"Oh yes. He kept a human arm in the freezer at his place in New York. I saw it."

Index ◆ ◆

bulletin boards (*cont.*)
ECHO, 135, 146–47, 150, 157
self-help on, 147–48
see also chat rooms; WELL
business:
and Internet, 231–33
Web sites of, 233–35, 240–41
WELL as, 186–88, 190–92, 193

Canine, Craig, 103
Carlyle, Thomas, 86, 148, 275
Carroll, Jon, 151, 169
Carson, Rachel, 273
censorship, 112; *see also* speech,
freedom of
Cerf, Vinton, 82, 231
chat rooms, 89, 137–45
anonymity of, 138
and associations, 227
avatars in, 260–61
on flaming and viruses, search for,
106–20
and getting attention, 225
hierarchies of, 107–8
lurking around, 104–5, 137, 168, 170
sex on, 139–45
spatial relationships in, 260–61
see also newsgroups
chug factor, 163
circuit-switched network, 79
Clipper Chip, 121–22
cluelessness, 173–74
Coate, John, 155, 156, 159, 179
commune, 157, 203–4
communication:
changed nature of, 77, 93
computer-to-computer, 80–81
via e-mail, *see* e-mail
media of, 82–83
community:
corporate, 240–41
destruction of, 150
face to face, 206–10
and home pages, 239

and management, 186–87, 259
nature of, 88
rules of, 195–96
virtual, 132–33, 148–49, 157–61,
197, 200, 208, 276, 277
WELL as, 191, 195–99
CompuServe, 84, 137, 150
and e-mail, 42, 43
and flaming, 102–3
Computers, Freedom and Privacy
Conference, 121–22
computers:
basic concepts of, 29–30, 69
bias against, 23
in boarding school, 19–23
breaking into, 147
communication between, 80–81
crashing, fear of, 31
games on, 20–23, 51, 132, 133–34,
139, 215
host, 80–81
hot potato routing in, 80
language of, 21–22
metaphors for, *see* metaphors
and solitude, 215, 252
upgrading of, 32–33
writing with, 29, 30–32, 47
see also technology
conferences, hosts of, 195, 201, 213,
245, 256–58
conferencing software, 149–51
connectivity, 252, 275
content:
control of, 85–86, 89, 103, 112,
186
encryption of, 121–22
copyright, and groupmind, 209
copyright infringement, 123
crackers and hackers, 146–47
crashing, fear of, 31
cyberspace:
early view of, 30
free speech in, 121; *see also* speech,
freedom of

fear (*cont.*)
  of groupmind, 179
  of Net growth, 191–92
  of posting a message, 171, 173, 177
  of worms, 100, 102, 124
Feltskog, E. N., 272
Figallo, Clifford, 155–56, 159, 181, 186
Fitzpatrick, Tom, 271
flame ghoul, 207
flaming, 83
  appeal of, 107
  article on, 119–20, 166–68
  and criticism, 180–81
  e-mail, 95, 97
  fear of, 118
  force of, 199
  and free speech, 96, 102–3, 112, 116, 172
  learning from, 180
  in newsgroups, 115, 116–18, 171
  opinions of, 103–6
  and piling on, 181
  reaction to, 95–100, 166–67, 171
  search for newsgroup on, 106–9, 171
  temptation of, 172–73, 174
  and worm, 95, 99–100, 125
Ford, Henry, 35, 65–66, 273
framework, value of, 127
freedom, *see* speech, freedom of
Frenchy, 19–23, 36, 215
frontier, 17
  of cyberspace, 14, 92, 130–31, 186, 208, 263
  freedom in, 130, 186, 276
  and Great Migration, 272
  loss of, 269
  and Manifest Destiny, 273
  metaphor of, 14, 92, 130–31, 263, 269
  mountain men of, 271–72
  Seabrook/Toomey families and, 92, 129–30, 262–69
Frontier in American History (Turner), 276

games, computer, 20–23, 51, 132, 133–134, 139, 215
Gans, David, 154
Gaskin, Stephen, 154–57, 179
Gates, Bill, 34–41
  as <billg>, 40, 47
  biographers of, 94–95, 274
  book by, 36, 250–51, 274, 277
  on competition, 57–58
  and DOS, 29
  e-mail from/to, 40–41, 43–45, 47–53, 54–60, 65–66, 93
  "face time" interview with, 38, 46, 60–65
  on literature, 58–59
  money and power and, 62, 75–76
  New Yorker story on, *see New Yorker*
  speaking engagement of, 250–51, 253–55
  as supersmart, 39
  and technology, 34–35, 251, 253, 260
Gilder, George, 33–34, 91, 114, 273–74
Gilmore, John, 85
Godwin, Mike, 170, 180
Gore, Albert, 53–54
government:
  antitrust actions of, 64
  and cyberspace, 53–54
  and networks, 83
  and privacy, 121–22
  and regulation, 121
graphical user interface (GUI), 28–29, 35, 136, 187
Greeley, Horace, 129
Greenleaf, William, 273
group hugs and beams, 185
groupmind, 148–53
  and copyright, 209
  covenant with, 197–98
  and flame article, 166–67
  and Katz, 187–95
  learning from, 180
  and many-to-many, 198
  mob behavior in, 179

and Over-Soul, 148–49, 179, 206
positive aspects of, 199–200
and pressure to conform, 197–99
problem solving with, 199
self-correcting mechanism of, 179, 183, 199
and sympathy, 200–203, 204–6
and troublemakers, 150
vulnerability of, 203
see also WELL
GUI (graphical user interface), 28–29, 35, 136, 187

hackers, and crackers, 146–47
Hacker's Conference, 153–54
Hall, Justin, 239
hard drive, 30
hate mail, *see* flaming
Hawthorne, Nathaniel, 277
Hegel, Georg Wilhelm Friedrich, 25
Henderson, Dan, 101–2, 120, 136
Herrero, Stephen, 275–76
Hirschfeld, Al, 277
home pages, 235–41
Horn, Stacy, 146–47
hot potato routing, 80
hypertext, 253

IAD (Internet Addiction Disorder), 217–226
IBM compatibles, and viruses, 124
icons, 28
idealism, 240
identity, nature of, 88
information highway, 14, 53, 54
information sharing, 82, 132
information theory, 79
intelligent agents, 39–40
interactive TV, 51–53
Internet:
addiction to (IAD), 217–26
business and, 231–33
chat rooms on, *see* chat rooms
control of content on, 85–86

and encryption, 121–22
free speech on, 85
grizzlies on, 274–75
history of, 78–84
lurking on, 104–5
as many-to-many medium, 85–90, 188
service providers for, 135–36, 187
use of term, 83–84
see also Usenet
Internet World, 220
IRL (in real life) friendships, 106, 164–166, 208
ISP (Internet service provider), 135–36, 187

Jefferson, Thomas, 71, 148, 276
Josephson, Matthew, 277
journalism:
vs. electronic posting, 176, 245
objectivity of, 48
and one-to-many media, 86–87, 275

Katz, Bruce, 186–95
and growing the WELL, 187–88
and social change, 190
WELL bought by, 161, 187
and WELL participation, 189
Keyworth, George, 274
Kleinrock, Leonard, 79, 80–81, 82
Kline, Charley, 81
Kolodny, Annette, 277
Kuhn, Robert E., 82

language:
acronyms and abbreviations, 113
computer, 21–22
harsh words, 199
networks based on, 113
new words, 113
for on-line postings, 176
Levy, Dan, 243–44, 246–49
Lunar Landing, 21–23
lurking, 104–5, 137, 168–71

Netscape, 161–62
networks:
  ARPAnet, 81–83
  breaking into, 147
  circuit-switched, 79
  control of content on, 85–86, 103, 186
  e-mail in, 82
  and groupmind, 148
  and hypertext, 253
  and language, 113
  many-to-many medium of, 80, 85–90, 188
  multiplex, 79
  network of, 83–84
  pressure to conform on, 197–99
  publishing on, 85
  spamming on, 111
  unsubscribing to, 225–26
  viruses spread by, 125
  weirdos on, 115–16
  writing style in, 83, 105
  see also Internet; Usenet
Neubarth, Michael, 219–20
newsgroups, 89
  and accountability, 150
  destruction of, 116, 118
  flaming in, 115, 116–18, 171
  hierarchies of, 107–8
  search for information on, 106–20
  surfing in, 109
  truth sought in, 115, 119–20
  see also chat rooms
New Yorker, the:
  Bill Gates story in, 37–40, 43, 72, 98, 169
  books previewed in, 273
  closings at, 261
  flaming responses to Gates article in, 95, 97–99
  Media Conference topic of, 167–183
  messages to Gates after publication of, 72–74
  messages to Seabrook after publication of, 74–78, 90–92, 95

"My First Flame" in, 119–20, 166–168
  writing assignments for, 174
nightclub metaphor, 157
Norstad, John, 120, 122–26
NSFnet, 83

on-line gathering places, 89; see also chat rooms
operating systems:
  DOS, 29, 35
  like medieval castles, 38
  Macintosh, 28–32
  Windows, 35
Optik, Phiber, 146–47
Oregon Trail, The (Feltskog), 272
Oregon Trail, The (Parkman), 17, 272–273
Over-Soul, 148–49, 179, 206

pagans, 109–11
Palace, The, 260–61
Parkman, Francis, Jr., 17, 272–73
pern, 151
point-to-point protocol (PPP), 135–36
posting:
  character and, 151
  fear of, 171, 173, 177
  games played in, 184–85
  vs. journalism, 176, 245
  vs. lurking, 173, 174–83, 184
  revelations in, 177
  scribbled, 205–6
  and threads, 196–97
power:
  and control, 36, 74
  and decentralization, 75
  Gates and, 62, 75–76
  Machiavelli and, 74–75
  of the nerds, 35–36
  and technology, 33, 34, 74–75, 96
  of virtual community, 208
PPP (point-to-point protocol), 135–36
print:
  articles sealed in, 171